DYNAMIC CHATGPT

AI STRATEGIES FOR SMALL BUSINESS

RON CLARK

KIM M. CLARK, M.S.B.

DEEP WATERS BOOKS

deepwatersbooks.com

Published by Deep Waters Books, PO Box 692301, Orlando, FL 32869 www.
deepwatersbooks.com

First Printing 2023
Printed in the United States of America

Portions of this book were generated by OpenAI (2023) ChatGPT (September 25 Version)
[Large language model] https://chat.openai.com.

Identifiers: ISBN: 978-1-956520-13-2 (hardcover) | 978-1-956520-14-9 (pbk.) | LCCN
2023950438

Publisher's Cataloging-in-Publication data

Names: Clark, Ron, 1964-, author. | Clark, Kim M., 1969-, author.
Title: Dynamic ChatGPT : AI strategies for small business / Ron Clark; Kim Clark, M.S.B.
Description: Orlando, FL: Deep Waters Books, 2023.
Identifiers: LCCN: 2023950438 | ISBN: 978-1-956520-14-9 (paperback) | 978-1-956520-13-
2 (hardcover)
Subjects: LCSH Artificial intelligence--Industrial applications. | Business enterprises--
Data processing. | Computational intelligence--Industrial applications. | Small business.
| Success in business. | BISAC COMPUTERS / Artificial Intelligence / General |
COMPUTERS / Business & Productivity Software / General | BUSINESS &
ECONOMICS / General | BUSINESS & ECONOMICS / Marketing / General |
BUSINESS & ECONOMICS / Business Communication / General | BUSINESS &
ECONOMICS / Small Business
Classification: LCC TA347.A78 C53 2023 | DDC 658.0563--dc23

CONTENTS

NOTE FROM THE AUTHORS

Before we created this book, we agreed to approach the subject of Artificial Intelligence and ChatGPT at a very basic level since this technology is both new and evolving. As business owners, we understand incorporating new tools into your business can be overwhelming. It can seem like you're just getting comfortable with one technology when another advancement takes its place.

We wrote this book as a primer, to introduce the foundations of ChatGPT by OpenAI. The content in each chapter is explained in depth to help you apply this knowledge to your specific business needs. To provide an additional understanding of the capabilities of ChatGPT and Artificial Intelligence, we created an entire online course program rather than writing a 700-page manual. Our online training programs can be accessed simply by visiting www.DynamicChatGPT.com.

Looking forward to partnering with you on your AI journey!

Ron and Kim

INTRODUCTION
[YOU REALLY SHOULD READ THIS]

Dynamic ChatGPT: AI Strategies for Small Business by Ron Clark and Kim M. Clark, M.S.B. serves as a guide for driven entrepreneurs, helping them harness the power of Artificial Intelligence (AI) through ChatGPT to streamline their operations, improve customer service, demystify marketing, and help businesses increase their revenues in an increasingly AI-driven world. Through practical examples of over one-hundred-fifty business prompts, twelve case studies, and several ethical considerations, small business owners will gain the knowledge and confidence to leverage AI.

Understanding that ChatGPT and AI's user interfaces are evolving at a rapid pace, some of the specific steps listed in this book may change over the course of time. If your business website does not have a significant web presence, ChatGPT will not be able to provide a specific response unless you include detailed information on your company and industry.

Also, if you require in-depth ChatGPT web and app integration, you may need to acquire the expertise of an AI professional. Lastly, if your organization has sensitive or proprietary data that needs to remain secure, educate all users on how to turn off chat history and treat ChatGPT as you would any other third-party software vendor.

Bravo on taking the first step to transform your business with AI!

GLOSSARY OF ARTIFICIAL INTELLIGENCE (AI) TERMS

IN AN EFFORT TO reduce your learning curve, we have included a helpful list of terms and definitions at the beginning of this book to kickstart your understanding of AI and its applications for small businesses.

1. **Algorithm:** An algorithm is a step-by-step procedure or set of rules for solving a specific problem or accomplishing a task.
2. **API (Application Programming Interface):** An API is a set of rules and protocols that allow different software applications to communicate with each other.
3. **Artificial Intelligence (AI):** AI refers to the simulation of human intelligence in machines that are programmed to think and learn like humans. It encompasses a range of technologies, including machine learning and natural language processing.
4. **Chatbot:** A chatbot is a computer program designed to simulate conversation with human users, often for customer support or information retrieval.

5. **Deep Learning:** Deep learning is a subset of machine learning that involves neural networks with many layers. It has been instrumental in advancing AI applications like image and speech recognition.
6. **Fine-Tuning:** Fine-tuning is the process of adjusting a pre-trained AI model to suit a specific task or application better.
7. **Machine Learning (ML):** Machine learning is a subset of AI that involves training algorithms to improve performance on a specific task by learning from data.
8. **Natural Language Processing (NLP):** NLP is a field of AI that focuses on the interaction between computers and human language. It enables machines to understand, interpret, and generate human language.
9. **Neural Network:** A neural network is a computational model inspired by the human brain, composed of interconnected nodes (neurons) that process and transmit information.
10. **Training Data:** Training data is the data set used to teach machine learning models. It consists of input data and corresponding output labels to help the model learn patterns and make predictions.

———

FREQUENTLY ASKED QUESTIONS (FAQS)

Before we begin, we wanted to address common questions about new technologies, especially when applying artificial intelligence to your business. Below are a few frequently asked questions and answers to assist you as you apply ChatGPT to your corporation.

What is the difference between AI and machine learning?

AI is a broader concept that encompasses the simulation of human intelligence in machines, while machine learning is a subset of AI that focuses on training algorithms to learn from data and improve their performance on specific tasks.

How can I get started with implementing AI in my small business?

Identify specific use cases where AI can add value, such as customer support, content generation, or data analysis. Research AI tools and platforms that are suitable for your needs and seek guidance from AI experts or consultants. Also, run the queries at the end of each chapter for your business and industry.

How can I ensure that AI implementations in my business are ethical and fair?

To ensure ethical AI usage, prioritize fairness, transparency, and data privacy. Establish ethical guidelines, regularly audit AI systems, and be aware of potential biases in data and algorithms.

Are there any regulatory considerations when using AI for small businesses?

Yes, it's important to stay informed about data privacy regulations that may apply to your business, such as GDPR in Europe or CCPA in California. Ensure that your AI practices comply with these regulations.

Is it expensive to integrate ChatGPT into my business operations?

The cost largely depends on your usage. As we will cover in chapter three, we opted to use the free version initially and then moved to the paid one once we understood ChatGPT's capabilities. For small businesses with moderate usage, the cost is quite manageable.

Is it safe to share business-sensitive information with ChatGPT?

It's always advised to avoid sharing any confidential or sensitive business information with ChatGPT or any other AI tool. While OpenAI has strict data privacy protocols, ensuring that data isn't stored or misused, it's a best practice to exercise caution.

By understanding and addressing these FAQs, small business owners can better grasp the potential and limitations of integrating ChatGPT into their operations. Now, let's get started!

CHAPTER 1
AI FOR SMALL BUSINESS SUCCESS

SMALL BUSINESS OWNERS only get paid when a product or service sells. Unfortunately, most small business owners wear a lot of hats, and most of those do not produce income. Whether you're a sole proprietor or have a large staff, you never have enough people, enough money, or enough time to grow your business.

That's why artificial intelligence (AI) has become the most powerful tool you can implement in your business today. AI increases productivity by automating and speeding up non-productive processes that can drain hours out of your day. Embracing and implementing AI technology in your business will allow you to compete against larger companies with more resources, thereby leveling the playing field and increasing your market share with minimal impact to your bottom line.

In an era where technological advancements are reshaping industries and markets at an unprecedented pace, small business owners find themselves uniquely positioned to harness the transformative power of AI to their advantage. This technology, once the domain of large corporations with substantial budgets, is now accessible to small businesses through tools like ChatGPT.

This book serves as the gateway to your journey into the world of AI for small business success, outlining the power of AI, the purpose and scope of this book, and introducing ChatGPT as your AI assistant.

THE POWER OF AI IN SMALL BUSINESS

The potential impact of AI on small businesses is immense. From streamlining operations to automating routine tasks, AI has the potential to transform the way small businesses operate. By leveraging the power of AI, small businesses can improve efficiency, reduce costs, and ultimately increase their bottom line. However, it's imperative to approach AI implementation cautiously to ensure it aligns with the overall business strategy. With the right approach, AI can be a game-changer for small businesses looking to stay competitive in today's rapidly evolving market.

Artificial Intelligence is revolutionizing how businesses operate globally, and small businesses are no exception. The power of AI lies in its ability to process vast amounts of data, automate tasks, make informed decisions, and enhance customer experiences. Here is a brief overview of how AI is transforming small businesses.

1. **Enhanced Customer Service:** One benefit of implementing specific strategies is improving customer service, resulting in better satisfaction, loyalty, and repeat business. AI-powered chatbots and virtual assistants like ChatGPT enable small companies to provide efficient, around-the-clock customer support. These AI solutions can answer customer inquiries, resolve common issues, and provide personalized recommendations, thereby improving customer satisfaction and retention.

2. **Data-Driven Decision-Making:** AI algorithms can analyze data faster and more accurately than humans. Small businesses can leverage AI to gain insights into customer behavior, market trends, and operational efficiencies. This data-driven decision-making approach can lead to better strategies and increased profitability.

3. **Marketing and Sales Optimization:** AI tools can help small businesses create targeted marketing campaigns, optimize advertising spend, and even predict customer preferences. ChatGPT can generate compelling marketing content, product descriptions, and email campaigns, saving time and resources.

4. **Automation of Repetitive Tasks:** Many small business owners are burdened by administrative tasks that consume their time and energy. AI can automate these tasks, such as appointment scheduling, invoicing, and inventory management, allowing entrepreneurs to focus on core business activities.

5. **Competitive Advantage:** Small businesses that embrace AI early gain a competitive edge. AI can help identify untapped opportunities, anticipate market changes, and stay ahead of the competition. It enables agility and adaptability in a fast-paced business environment.

PURPOSE AND SCOPE OF THE BOOK

The purpose of this book is to demystify AI and empower small business owners with practical knowledge and actionable insights. Whether you are a restaurant owner, an e-commerce entrepreneur, a consultant, or any other type of small business owner, this book will guide you on how to leverage AI to achieve your business goals. Here's what you can expect from this book:

1. **Understanding AI for Small Businesses:** AI can be intimidating. This book will explain the fundamentals of AI in a non-technical way so you can grasp the concepts without any prior technical expertise.

2. **Practical Applications and Case Studies:** You will learn how AI can be applied across various aspects of your business, including customer service, marketing, data analysis, and automation.

3. **Hands-On Guidance:** Step-by-step instructions and real-world examples will illustrate how to implement AI solutions, including the use of ChatGPT, in your small business.

4. **"Ask the Bot" Business Prompts:** At the end of each chapter, we included five ChatGPT prompts for you to apply the knowledge you've learned in your specific industry, market, and company. Query tips are included as well.

5. **Ethical Considerations:** We will emphasize the importance of ethical AI usage, including data privacy and responsible practices, to maintain trust with your customers and stakeholders.

6. **Future Trends:** We will discuss emerging AI trends and technologies that may impact small businesses in the near future, helping you stay ahead of the curve.

ABOUT CHATGPT: YOUR AI ASSISTANT

ChatGPT is a powerful AI language model developed by OpenAI. It has been trained on vast amounts of text from the internet, making it capable of understanding and generating human-like text responses. ChatGPT is designed to assist users in a wide range of tasks, from answering questions and developing content to providing recommendations and automating conversations. Below is a short list of ChatGPT's key features and applications.

- **Natural Language Understanding:** ChatGPT can understand and conversationally respond to a text, making interactions with AI more intuitive.
- **Customization:** You can customize ChatGPT to align with your business's tone, style, and objectives, creating a more personalized AI assistant.
- **Scalability:** ChatGPT can handle multiple customer inquiries simultaneously, providing efficient and consistent support.

- **24/7 Availability:** AI doesn't need sleep, breaks, or vacations, so ChatGPT can provide assistance around the clock, enhancing customer service.
- **Customer Support:** ChatGPT can handle common customer inquiries, freeing up human resources for more complex tasks.
- **Content Generation:** It can generate blog posts, product descriptions, social media content, and more, saving time required for content creation.
- **Data Analysis:** ChatGPT can assist in analyzing customer feedback, market trends, and competitor data, enabling data-driven decision-making.
- **Automation:** You can automate appointment scheduling, order processing, and other administrative tasks with ChatGPT.

In the following chapters, we will dive deeper into how ChatGPT can be integrated into your small business operations to improve efficiency, customer satisfaction, and, ultimately, your bottom line. As we embark on this journey into the world of AI for small business success, remember that AI, including ChatGPT, is an incredibly powerful tool at your disposal to help you achieve your business goals, innovate, and thrive in a rapidly evolving business landscape.

CASE STUDY: TRANSFORMING OPERATIONS WITH AI

As business owners, we found that we learn best by first seeing and then doing. In each chapter, we'll dive into case studies that showcase the successful implementation of ChatGPT and other AI technologies in small businesses. These success stories demonstrate how AI can drive growth, improve efficiency, and transform various aspects of business operations. We have intentionally not mentioned the names of the companies to protect their privacy. The first one we will review is a small urban landscaping business.

This horticulture company specializes in creating sustainable and aesthetically pleasing green spaces in urban areas. While they had a strong vision and customer base, their operational efficiency was hampered by manual processes and a lack of technological integration.

Challenge: Their ability to scale and respond to market demands quickly was hampered by issues with project management, customer communications, and resource allocation. They wanted to avoid these obstacles by improving operational efficiency and reducing costs without losing their personal touch with clients.

Solution: Leveraging AI technologies to streamline their project management, customer service, and market analysis, as well as automate administrative tasks. The prompts used and outcomes are below.

Prompt: Generate a project timeline and resource allocation plan for an urban landscaping project.

Outcome: AI algorithms predicted project durations and resource needs based on past data, which helped in planning, scheduling, and handling routine customer inquiries on their website, freeing up time for staff to focus on more complex queries.

Prompt: Respond to customer queries regarding service availability and booking procedures.

Outcome: Integrated a chatbot on their website for instant responses to common questions.

Prompt: Analyze current trends in urban landscaping and customer preferences based on social media data.

Outcome: AI analyzed data from various sources to identify emerging trends in urban landscaping.

Prompt: Automate monthly invoicing and payroll processing for a small landscaping business.

> Outcome: AI software streamlined back-office operations, leading to a reduction in paperwork and manual errors.

Results: Six months after AI implementation, the landscaping company experienced significant improvements in operations, customer satisfaction, and cost savings.

- **Increased Operational Efficiency:** Project turnaround time improved by 31.8%, with more efficient resource utilization.
- **Enhanced Customer Satisfaction:** The response time for customer inquiries was reduced by 74.3%, leading to higher customer satisfaction rates.
- **Cost Savings:** Operational costs were reduced by 22.4% due to the automation of routine administrative tasks.
- **Informed Business Decisions:** Better market insights led to a more customer-focused service offering, which increased new business acquisitions by 18%.

Conclusion: This landscaping company's utilization of AI showcases the significant impact ChatGPT can have on a small business. By strategically integrating AI into their operations, they improved efficiency, reduced costs, and stayed competitive in a rapidly evolving market. This case study illustrates that with careful planning and the right approach, AI can be a game-changer for small businesses aiming to optimize their operations and grow their bottom line.

ASK THE BOT: ADDITIONAL AI COMMANDS

To continue to assist you in applying the knowledge you learned from this chapter, open up and log into ChatGPT on your mobile device. Go to the Apple App Store or Google Play Store and download the "ChatGPT" app powered by OpenAI or on your computer by visiting www.chat.openai.com. Detailed instructions on how to set up your account can be found in chapter three of this book. Then, enter the queries below into ChatGPT.

Important note: In the "Ask the Bot" section at the end of each chapter, you will enter your company name, info, website, and industry. If you have a minimal or nonexistent web presence, ChatGPT will not provide a specific answer. In these instances, enter particular information on your business to receive an accurate response.

> How do you customize ChatGPT for [enter your business, website, or industry]?

> Create a marketing campaign for [enter your business, website, or industry].

> Create twenty hashtags for this [enter your business, website, or industry] marketing campaign.

> What administrative tasks can ChatGPT do for [enter your business, website, or industry]?

> What anticipated market changes and competitive edge should [enter your business, website, or industry] be aware of?

These prompts are just a few examples of how AI can help small business owners and operators better understand ChatGPT and how it can be strategically utilized to benefit their specific needs and goals.

ADDITIONAL TRAINING

Since ChatGPT is continually evolving and each industry is unique, we have created extensive online training programs on our website, www.DynamicChatGPT.com, to further assist you in harnessing the power of AI for your small business.

CHAPTER 2
UNDERSTANDING CHATGPT

IMAGINE RUNNING into a debilitating technology issue in your business that you have no idea how to fix. You pick up the phone and call Elon Musk. In a few moments, you receive the much-needed solution to your problem and your company is saved. Or what if you're not sure how to create a compelling advertising campaign for your new product or service offering and if you should place the ads on Google or Facebook? Suddenly, Sergei Brin (the co-founder of Google) calls your cell phone and Mark Zuckerberg (Facebook founder) texts you, offering unique insights on placements for your industry and target markets.

This is what ChatGPT is like. The combined experience of millions of people united in one app gives you instantaneous 24/7 access from your computer, tablet, or smartphone. As a business owner, understanding how and why AI works can make the difference between your operations thriving or dying.

In this chapter, we will delve into a comprehensive understanding of ChatGPT, a powerful AI language model developed by OpenAI. We will explore ChatGPT, how it works, its key features and capabilities, and the tangible benefits it offers small businesses.

WHAT IS CHATGPT?

ChatGPT is a state-of-the-art artificial intelligence language model designed to engage in human-like conversations. It is part of a broader family of AI models known as the GPT (Generative Pre-trained Transformer) series. At its core, ChatGPT is a deep neural network that has been pre-trained on a vast corpus of text from the internet, allowing it to understand and generate text in a remarkably human-like manner.

CHARACTERISTICS OF AI

Artificial Intelligence (AI) simulates human intelligence in machines, enabling them to think, learn, and solve problems. AI systems are developed for tasks such as natural language processing, pattern recognition, decision-making, and complex problem-solving to create technology that functions intelligently and improves over time. Below is a summary of the key characteristics of AI.

1. Generative: AI is built on a generative model, which means it can produce new, coherent responses based on the input it receives. This allows it to craft contextually relevant responses in real-time conversations. The generative nature of ChatGPT makes it versatile for many tasks, from answering questions to creative writing.

- **How It Works:** Instead of relying on predefined templates or databases of responses, ChatGPT uses a neural network to dynamically generate answers based on the patterns it has learned from vast amounts of textual data.
- **Implications:** This dynamic generation allows for unique, real-time responses that can cater to an endless variety of user inputs. However, it might occasionally produce unexpected or only partially accurate outputs.

2. Language Understanding: AI possesses advanced natural understanding capabilities, enabling it to comprehend and generate human-like text in conversations. This allows for intuitive interactions, as the model can grasp context, nuances, and intent within textual exchanges. As a result, ChatGPT provides users with coherent and contextually relevant responses in a wide array of topics and scenarios.

- **Depth of Understanding:** ChatGPT doesn't just match keywords or phrases; it evaluates the structure and context of the input to generate meaningful responses.
- **Implications:** This nuanced understanding allows for more natural conversations and reduces misunderstandings. Yet, it's essential to remember that every model is flawed, and there might still be occasional hiccups in comprehension.

3. Adaptability: ChatGPT's AI adaptability is a huge advantage for small businesses, as it can be fine-tuned to cater to specific industry needs and customer interactions. Its versatility means it can adjust to different business domains, ensuring relevant and accurate responses. As a result, small businesses can leverage ChatGPT to enhance customer support, generate content, and automate various tasks with minimal setup.

- **Customization:** Beyond its base training, ChatGPT can be further trained on specialized data sets to cater to specific industries or niches, such as medical, legal, or entertainment sectors.
- **Implications:** This adaptability means businesses can mold the model to their unique needs. However, the quality of the fine-tuning data is a necessity, and care must be taken to avoid biases or inaccuracies.

4. Scalability: ChatGPT's scalability is essential for small businesses, allowing them to handle a wide range of interactions without requiring extensive infrastructure. Whether dealing with a handful of inquiries or managing a surge in customer engagement, ChatGPT can adjust seamlessly. This ensures consistent and efficient customer service, regardless of the volume, without additional costs.

- **How It Works:** ChatGPT, built on advanced neural network architectures, can manage concise interactions and longer, more intricate dialogues.
- **Implications:** This scalability ensures that the model remains efficient and effective regardless of the conversation's depth or length. It can be employed in contexts ranging from quick customer support chats to in-depth discussions or brainstorming sessions.

In summary, the capabilities of ChatGPT position it as a transformative tool in the realm of conversational AI. Its generative nature, language understanding, adaptability, and scalability make it a valuable asset across various applications. Nevertheless, ongoing refinement and knowledge of its limitations are important to harness its full potential responsibly.

HOW CHATGPT WORKS

ChatGPT's operation is rooted in a transformer architecture, which is a type of neural network that excels in capturing sequential data patterns, such as those found in natural language. Here's a simplified overview of how ChatGPT relates to human development.

Pre-training (Learning Basics): This stage is akin to early childhood in humans, where basic language skills are developed. During the pre-training phase, ChatGPT is exposed to an extensive data set comprising text from books, articles, websites, and more. It learns to predict the next word in a sentence, thereby acquiring an understanding of grammar, syntax, and contextual relationships.

Just as a child learns words, grammar, and sentence structure through exposure to language in their environment (like conversations, books, and media), ChatGPT, during pre-training, learns language fundamentals from a vast data set. It's like a child making sense of the language rules and patterns around them.

Fine-tuning (Specialized Learning): This resembles adolescence and early adulthood in human development, where individuals receive more specialized education or training. Following pre-training,

ChatGPT can be adjusted for specific tasks or domains. This necessitates exposing the model to a narrower data set carefully curated to align with the intended application. Fine-tuning helps customize ChatGPT for particular use cases, such as customer support or content generation.

In humans, this is the phase of deepening knowledge in specific areas and acquiring skills tailored to particular interests or future professions. Similarly, ChatGPT undergoes specific calibrations with distinct data sets to specialize in certain tasks or domains, akin to a student choosing a major or a career path and focusing their studies accordingly.

Inference (Application of Knowledge): This is comparable to adulthood, where individuals apply what they've learned in real-world scenarios. Inference is the process where ChatGPT takes a text input and generates a text response. The model utilizes the knowledge gained during the pre-training and fine-tuning stages to create coherent and contextually appropriate responses.

In this phase, ChatGPT uses its acquired and fine-tuned knowledge to generate responses to new inputs, much like an adult applying their education, experiences, and life lessons to make decisions or solve problems in their personal and professional life.

Contextual Understanding (Social Intelligence): This application mirrors the development of social intelligence in humans. ChatGPT maintains context throughout a conversation, providing consistent and relevant responses to previous interactions. It can reference prior messages to understand the current conversation's requests.

Just as humans learn to understand and respond to social cues and maintain context in conversations over time, ChatGPT keeps track of the input and output history to provide relevant and coherent responses. It's akin to an individual recalling previous discussions and using that context to inform their current interactions.

In summary, the stages of ChatGPT's operation reflect a journey from learning basic skills to applying them in contextually aware and socially intelligent ways, paralleling human growth from infancy through adulthood.

KEY FEATURES AND CAPABILITIES

ChatGPT is a versatile AI tool that brings a range of powerful features and capabilities to the table. These attributes make it an ideal solution for small businesses aiming to optimize operations and enhance customer interactions.

Natural Language Understanding: ChatGPT boasts natural language understanding, allowing it to interpret and generate text in a human-like manner. This means that interactions with ChatGPT are intuitive, engaging, and user-friendly. Whether you're communicating with customers or seeking insights from the AI assistant for e-commerce solutions, the natural language processing capabilities ensure seamless communication.

Customization: Small businesses can tailor ChatGPT to align with their brand's unique tone, style, and specific objectives. Customization options empower businesses to create a more personalized and coherent AI assistant. By adapting ChatGPT to your brand's identity, you can ensure customer interactions reflect your company's values and message.

Multilingual Support: In today's globalized world, catering to diverse audiences is pivotal. ChatGPT offers multilingual support, enabling businesses to communicate with customers worldwide. Whether you're providing customer support, marketing content, or product information, ChatGPT's multilingual capabilities ensure that language is not a barrier to engagement.

24/7 Availability: ChatGPT's 24/7 availability is a game-changer for small businesses looking to enhance customer support and responsiveness. With around-the-clock assistance, you can promptly address customer inquiries and concerns, improving overall customer satisfaction and loyalty.

Scalability: Small businesses often face fluctuations in customer inquiries, with peak hours demanding extra attention. ChatGPT's scalability comes to the rescue by efficiently handling multiple customer interactions simultaneously. This ensures your business can provide consistent and efficient customer service, even during high-demand periods.

Content Generation: This is a time-consuming task for many small businesses. ChatGPT streamlines this process by generating various types of content, such as blog posts, product descriptions, and marketing copy. This feature saves valuable time and resources, allowing businesses to focus on other critical aspects of their operations.

By harnessing these key features and capabilities, small businesses can leverage ChatGPT as a versatile AI assistant to improve customer interactions, streamline content creation, and enhance their overall operations. ChatGPT's natural language understanding, customization options, multilingual support, 24/7 availability, scalability, and content generation capabilities make it an invaluable tool in the modern business landscape.

BENEFITS FOR SMALL BUSINESSES

Integrating ChatGPT into your small business operations brings forth many advantages that can substantially impact your efficiency, customer satisfaction, and overall success. Let's delve into some of these tangible benefits.

Enhanced Customer Support: Small businesses often struggle to provide timely and efficient customer support due to limited resources. ChatGPT steps in as a reliable assistant, capable of handling customer inquiries swiftly and accurately. By ensuring quick responses to customer queries, ChatGPT contributes significantly to improved customer satisfaction and retention rates. Happy customers are more likely to become loyal customers and share their positive experiences.

Time and Resource Savings: Time and resources are invaluable commodities for small businesses. ChatGPT contributes to resource optimization by automating routine tasks such as appointment scheduling, data analysis, and so much more. This automation frees up human resources to focus on higher-value and more productive activities, such as strategic planning, innovation, and creative tasks that require human ingenuity.

Cost Efficiency: Small businesses often operate on tight budgets, prioritizing cost efficiency. ChatGPT reduces the need for additional customer support agents or content creators, as it can handle tasks traditionally performed by humans. This translates to significant cost savings over time, allowing small businesses to allocate their budgets more effectively.

Consistency: Maintaining consistent and standardized responses is foundational for building trust and delivering a seamless customer experience. ChatGPT excels in this regard, providing uniform assistance to customers regardless of the time of day or the volume of inquiries. Consistency in customer interactions enhances your brand's reliability and professionalism.

Data-Driven Insights: Small businesses can harness ChatGPT's capabilities to extract valuable insights from customer feedback, market trends, and competitor data. By automating the data analysis process, ChatGPT empowers you to make data-driven decisions and formulate more informed strategies. These insights can be pivotal in adapting to market changes and staying ahead of the competition.

Competitive Advantage: Staying ahead of the curve is vital in a competitive business landscape. Small businesses that embrace AI, like ChatGPT, gain a competitive edge by becoming more responsive, efficient, and innovative. This advantage enables them to stand out in their respective markets, attract more customers, and adapt to evolving customer preferences and industry trends.

Armed with this knowledge and the tools at your disposal, you'll be well-equipped to unlock the full potential of AI and ChatGPT to drive your small business to new heights of success. By embracing the benefits of AI, small businesses can not only survive but thrive in today's dynamic business environment.

CASE STUDY: BOOSTING REVENUE WITH TARGETED CHATGPT PROMPTS

A small artisan bakery known for its handcrafted pastries and custom cakes is facing stiff competition from larger chains and local bakeries. Despite a loyal customer base and exceptional product quality, they struggled to increase revenue and market share. They also realized the need for innovative marketing and customer engagement strategies to enhance their business's visibility and sales.

Challenge: The bakery's primary concerns were limited resources and time constraints, common in small businesses. They sought a cost-effective, efficient way to enhance customer engagement, personalize marketing efforts, and streamline administrative tasks without overburdening their small team.

Solution: Using specific AI prompts, they tailored ChatGPT's capabilities to the bakery's business marketing, customer interaction, administrative, and product development needs. The prompts used and outcomes are below.

Prompt: Create a monthly marketing campaign for a bakery specializing in custom cakes and pastries.

Outcome: ChatGPT generated unique monthly themes, aligning with local events and seasonal trends, and crafted engaging social media posts and email newsletters.

Prompt: Draft responses for common customer inquiries about custom cake orders, including pricing, flavors, and delivery options.

Outcome: ChatGPT provided instant, consistent, and informative replies, improving customer satisfaction and freeing staff time.

Prompt: Analyze customer feedback and suggest three new pastry flavors for the fall season.

Outcome: ChatGPT identified popular flavors and trends from customer feedback data, creating well-received seasonal offerings.

Prompt: Organize a weekly schedule for staff shifts, baking, and delivery for a small bakery.

Outcome: ChatGPT helped streamline scheduling, ensuring optimal staff allocation and operational efficiency.

Prompt: Create twenty hashtags for promoting a local bakery's seasonal pumpkin spice products.

Outcome: ChatGPT provided trendy, relevant hashtags, increasing social media visibility and engagement.

Results: Within six months of implementing ChatGPT, the bakery experienced significant improvements, such as increased revenue, improved customer engagement, and operational efficiency.

- **Revenue Increase:** They saw a 24.8% increase in revenue, attributed to more effective marketing campaigns and an expanded customer base.
- **Customer Engagement:** Social media engagement soared by 42%, significantly increasing followers, shares, and likes.
- **Operational Efficiency:** Time spent on customer inquiries was reduced by 37.4%, allowing the team to focus more on product development and customer service.
- **Product Innovation:** The introduction of AI-suggested pastry flavors resulted in a 17.9% increase in sales of seasonal items.

Conclusion: This case study is a testament to the power of AI in transforming small business operations. By leveraging specific prompts tailored to their industry, the bakery harnessed ChatGPT's capabilities effectively, leading to significant revenue growth, enhanced customer engagement, and improved operational efficiency. This case study demonstrates that even small businesses with limited resources can reap substantial benefits from AI technology when used strategically.

ASK THE BOT: ADDITIONAL AI COMMANDS

To further assist you in applying the knowledge you learned from this chapter, open up and log into ChatGPT on your mobile device or computer and enter the following queries.

> Provide an overview of what ChatGPT is and how it works for [enter your business, website, or industry].
>
> List practical ways and examples in which ChatGPT can improve operations and customer interactions in [enter your business, website, or industry].
>
> How can I customize ChatGPT to align with [enter your business, website, or industry]'s tone, style, and objectives?
>
> Are there best practices for tailoring ChatGPT to [enter your business, website, or industry]?
>
> What steps are involved in implementing ChatGPT into my website, customer support, or other systems for [enter your business, website, or industry]?

These prompts can help you explore various ways ChatGPT can be leveraged to enhance small business operations, from customer support to operations management.

ADDITIONAL TRAINING

Since ChatGPT is continually evolving and each industry is unique, we have created extensive online training programs on our website, www.DynamicChatGPT.com, to further assist you in harnessing the power of AI for your small business.

CHAPTER 3
GETTING STARTED WITH CHATGPT

THERE IS an old saying that you can't teach old dogs new tricks, but that's not true. We adopted our yellow lab at fifteen months. She was housebroken and knew how to sit. Over a few months, we taught her over two dozen verbal and hand commands. She's now six years old and well-trained. She still loves learning new skills and is continually eager to please. Some of her natural aptitude for learning comes from her breed, but without the proper knowledge and consistent training, her giftings would remain untapped, just like ChatGPT for business owners.

If you don't utilize the immense capabilities of this incredible resource, you are hamstringing your corporation. It's like someone gave you a powerful computer, and you leave it in the box unopened while you continue to use antiquated processes, producing diminishing results.

As consultants, we tell entrepreneurs all the time if you aren't learning something new every day, your business is declining. One hundred years ago, you could get away with minimal abilities and experience and run a business for decades without ever learning any additional expertise. Not anymore. Technology is moving at light speed and rapidly changing the corporate landscape. You will be out of business if you're not keeping up with the technology.

We understand that AI might seem like a foreign language, which is why we wrote this book and provided additional online training. As an entrepreneur, it won't take you long to understand and implement it in your company. In this chapter, we will guide you through the steps to get started with ChatGPT. Whether you're a small business owner looking to implement AI-powered customer support or leverage ChatGPT's capabilities for content generation, understanding how to set up, navigate, and customize your ChatGPT account is vital to your success. Below are step-by-step instructions on how to start this powerful toolset for your organization.

SETTING UP YOUR CHATGPT WEB ACCOUNT

Getting started with ChatGPT is a straightforward process. OpenAI has designed an intuitive platform that allows you to create an account and access the AI model. Here's a step-by-step guide to setting up your ChatGPT account on your internet browser. The mobile app is slightly different; we'll cover those instructions next.

1. **Visit the ChatGPT Website:** Visit chat.openai.com.
2. **Sign Up:** Click the "Sign Up" or "Get Started" button to initiate the registration process. You will need to provide your email address and create a password for your account.
3. **Verification:** Verify your email address by clicking the link sent to your inbox. This step is fundamental for account security and access to ChatGPT.
4. **Payment Information:** Depending on the pricing model OpenAI offers at registration, you may need to enter your payment information. Review the options carefully and choose the one that best suits your needs. We used the free version for quite some time and then upgraded to the paid subscription service. The robust paid version better fits our growing and comprehensive needs.
5. **Access Your Dashboard:** Once your account is set up and verified, you can log in to your ChatGPT dashboard. Here, you can access the AI model and its various features.

SETTING UP YOUR CHATGPT MOBILE APP ACCOUNT

Creating your ChatGPT account on the mobile app is a straightforward process that enables you to access AI-powered assistance on the go. Here's a quick guide to get you started.

1. **Access the ChatGPT Mobile App:** Begin by downloading the ChatGPT mobile app from your device's app store. It's available for both iOS and Android platforms.
2. **Create Your Account:** Launch the app and follow the on-screen prompts to create your account. You'll need to provide your email address and choose a strong password.
3. **Verification:** Some apps may require email verification or a one-time verification code sent to your email. Follow the instructions to confirm your account.
4. **Set Preferences:** You can personalize your ChatGPT experience by setting preferences such as language, communication style, and specific business objectives.
5. **Explore the Interface:** Familiarize yourself with the mobile app's user interface. You'll typically find a chat window where you can interact with ChatGPT, as well as options to customize settings and access additional features.
6. **Start Conversations:** With your account set up, you can start using ChatGPT on your mobile device. Simply enter a prompt to initiate a conversation, ask questions, seek assistance, or explore the various use cases discussed in this guide.

Setting up your ChatGPT account on the internet and mobile app empowers you to access AI assistance at all times, making it a valuable tool for small business owners on the move and at the office. Enjoy the convenience of real-time support and data-driven insights right at your fingertips.

EFFECTIVELY NAVIGATING THE CHATGPT WEB INTERFACE

Navigating the ChatGPT interface is a straightforward and user-friendly experience designed to ensure your interactions with the platform are seamless and intuitive. To maximize your effectiveness in using ChatGPT, it's important to familiarize yourself with the various elements and functionalities of the interface. Let's explore how you can navigate your ChatGPT interface on the internet effectively to make the most out of your AI assistant.

- **Dashboard Overview:** After logging in, you will land on the dashboard. You can see your recent conversations, access settings, and initiate new discussions.
- **Starting a Conversation:** To create a conversation, click the "New Conversation" button. You can enter your initial message or query in the chatbox. The more specific you are in asking your questions, the better responses you will receive.
- **Interactive Chat:** ChatGPT will respond to your message in real-time, generating human-like responses. The conversation will be displayed in a chat-like format, making it easy to follow. You can then click on the clipboard icon on the right of the response to copy the answer. Clicking on the "thumbs-up" or "thumbs-down" icon refines and optimizes the chatbot's performance and accuracy.
- **Context Management:** ChatGPT maintains context within the conversation, which means it remembers the previous messages and responds accordingly. This helps in having coherent and context-aware interactions. You can also return to a "conversation" and enter additional requests for further information. For example, you might enter, "Create digital ad copy for Google ads for a new and improved widget with the following features…" Then, later, you might return to that conversation and ask it to create Facebook ad copy and twenty supporting social media hashtags.

- **User Commands:** ChatGPT understands various user commands. For instance, you can use "/reset" to clear the conversation history or "/help" to get additional information.
- **Access to Documents:** You can easily access and share documents with ChatGPT during a conversation. This feature can be handy for collaborative tasks or when you need AI-generated content based on specific text.

EFFECTIVELY NAVIGATING THE CHATGPT MOBILE APP INTERFACE

Navigating the ChatGPT mobile app interface is designed to give you a streamlined and optimized user experience on your mobile device. With touches and swipes, you can interact with your AI assistant smoothly. Below, we'll guide you through the core functionalities of the ChatGPT mobile app interface, ensuring that you can make the most of your AI-powered companion on the go.

- **Mobile Dashboard:** Upon opening the app, you're greeted with the mobile dashboard. Here, you can view a list of your recent conversations, jump back into ongoing discussions, and navigate to the app's settings. A floating action button, usually represented by a "+" icon, is your pathway to initiating new chats.
- **Initiating Conversations:** Tap the "+" or "New Conversation" icon. You'll be presented with a chat interface. Type in your message or query. As always, specificity helps ChatGPT provide you with the best possible answers.

> Reminder: When prompted to "Ask the Bot" for specific industry-based queries, remember to provide essential details about your business, especially if it lacks a prominent web presence.

- **Chat Interaction:** Responses from ChatGPT are designed for the mobile screen and appear in a familiar messaging format. Beside each response, you might find icons:

28

Clipboard icon: Tap to copy the response.

Thumbs-up/thumbs-down: This is where you can influence and provide feedback on the AI's response to improve future interactions with your queries.

- **Contextual Conversations:** Like its web counterpart, the mobile ChatGPT retains context throughout your conversation. This ensures relevant and context-aware responses. If you start a conversation about a particular topic, you can easily return to it later and pick up where you left off.
- **Gesture Commands:** The mobile interface incorporates gesture-based commands. A simple swipe downwards can refresh the chat or clear any input. Swipe left to go back to the dashboard. Additionally, typing commands like "/reset" or "/help" in the chatbox provides similar functionality as on the web platform.
- **Document Sharing:** Even on mobile, ChatGPT is ready to handle document-based tasks. You can upload and share documents directly in the chat through the attachment icon (often represented by a paperclip or similar). This is useful when seeking AI insights or content based on specific documents.

In conclusion, while the ChatGPT mobile app is tailored for a handheld experience, it retains the power and versatility of its web-based version. Familiarize yourself with these functionalities, and your on-the-go AI interactions will be both efficient and enjoyable.

CUSTOMIZING WEB CHATGPT FOR YOUR BUSINESS

Customizing ChatGPT for your business involves tailoring the AI to match your unique brand tone, style, and objectives. This customization is critical in ensuring ChatGPT aligns with your business's specific needs and communicates effectively with your audience. OpenAI offers a variety of tools and options on its website to assist in this process. These tools enable you to modify ChatGPT to become a more personalized AI assistant, ensuring it resonates with your business's ethos and enhances customer interactions. By leveraging these customization options, you can create a ChatGPT model that not only understands and responds to queries effectively but also embodies the voice and personality of your brand.

ChatGPT Settings

Understanding that ChatGPT and AI's user interfaces are evolving at a rapid pace, some of the specific steps may change over time. As of the time of print, navigate to "My GPTs" under the "Explore" tab of your profile in the bottom left to create a customizable version of ChatGPT for your specific purposes.

- **System Message:** You can set a system message to provide context to ChatGPT about its role in the conversation. For example, enter the prompt, "You are a customer support assistant."
- **User Message:** You can specify instructions or expectations for the AI by sending a user message at the beginning of the conversation.

Style and Tone: Again, at the time of print, click on your name at the bottom left and select "Custom Instructions." Here, you can modify the style and tone of ChatGPT's responses to match your corporate brand identity. Suppose your business is known for being professional and formal. In that case, instruct ChatGPT to use a similar tone in its response. You would indicate your selection in the "What would you like ChatGPT to know about you to provide better responses?" and "How would you like ChatGPT to respond?" fields under the "Custom Instructions" setting box.

Use of Specific Data: If your business has proprietary data or knowledge that you want ChatGPT to use in responses, you can provide that information during the conversation. ChatGPT can reference specific data points to provide accurate and customized answers. If your business operates in a specialized industry like finance, healthcare, or law, you can provide ChatGPT with data and information pertinent to these fields. For example, a financial services company can equip ChatGPT with the latest market trends, investment strategies, or regulatory information to ensure it provides accurate financial advice to customers.

Domain-Specific Knowledge: If your business operates in a niche industry or domain, you can fine-tune ChatGPT to have domain-specific knowledge. This involves training the AI model on a data set relevant to your industry, ensuring it can provide more accurate information and insights. For example, a winery wants to use ChatGPT to educate customers about wine varieties, pairings, and their winemaking process, as well as to assist with purchases. After entering their wine descriptions, food pairings, winemaking process, and vineyard history, customers who visit their website can get expert advice on wine selections and food pairing suggestions. The possibilities are endless.

Iterative Improvement: ChatGPT continuously fine-tunes and customizes responses based on your user feedback and evolving business needs. Regularly revisiting and refining your AI settings can lead to better performance and alignment with your objectives.

As an example, an online handmade personalized gift store wants to use ChatGPT to handle customer inquiries, provide product recommendations, and assist with order-related queries. After they entered their information about their product range, ordering process, shipping policies, and return guidelines in ChatGPT, as well as customer feedback, ChatGPT identifies patterns, such as frequent queries about certain products or customer issues. This allows ChatGPT to handle common customer inquiries effectively, offer better guidance on order-related issues, and even suggest personalized gift recommendations based on customer preferences or past purchases. Over time, through this iterative process, ChatGPT becomes more aligned with the company's specific needs and queries.

CUSTOMIZING THE CHATGPT MOBILE APP FOR YOUR BUSINESS

As of the last update in January 2022, OpenAI hadn't released any specific functionality for customizing the ChatGPT experience for business use within a mobile app context. The primary interface of ChatGPT doesn't inherently offer business-specific customization on the app or the web. Instead, it's a generalized tool.

However, if you wanted to integrate ChatGPT's capabilities into your business's mobile app, you would typically follow these steps.

- **API Integration:** OpenAI provides an API for GPT-based models. Developers can use this API to integrate ChatGPT into mobile apps. This way, you can customize the experience based on your business needs.
- **Branding & UI Design:** Once integrated, you can style the chat interface to match your business's brand identity, including logos, colors, and fonts.
- **Training & Tuning:** While GPT models like ChatGPT come pre-trained, there are ways to fine-tune them on specific data sets. If OpenAI permits, you can fine-tune a model based on your business data to make it more specialized for your use. This would require adherence to OpenAI's guidelines and may involve additional costs.
- **User Guidance:** To ensure that users get the most accurate and relevant responses, you might provide guidelines or prompts within the app, suggesting how to phrase queries or what kind of questions they can ask.
- **Feedback Mechanism:** Implement a feedback system where users can rate or comment on the AI's responses. This can be useful for continual improvement and understanding areas where AI might be lacking.
- **Data Privacy & Security:** Ensure the integration respects user privacy, especially if the conversations contain sensitive business or personal information. Clearly communicate to users how their data will be used and stored.

Again, if your company uses and has access to sensitive or proprietary data, make sure you instruct all users to turn off chat history and treat ChatGPT and all generative AI as you would any other third-party software. We also suggest regularly reviewing generative AI security protocols to protect your business data.

- **Continuous Monitoring:** AI can sometimes produce unexpected or undesirable outputs. It's so important to monitor interactions and make adjustments as needed to maintain a high-quality user experience.

Recommendation: Always check for the latest from OpenAI's official website or communication channels to see if there have been new releases or updates related to business integrations or mobile app customizations.

In the following chapters, we will explore the practical applications of ChatGPT in different aspects of small business operations, including customer support, content generation, data analysis, and more. Understanding how to set up and customize your ChatGPT account is the first step toward harnessing the power of AI to enhance your business's efficiency and customer engagement.

CASE STUDY: TAILORING CHATGPT FOR PERSONALIZED CUSTOMER INTERACTION

A mid-sized travel agency known for its attention to detail and customer-centric approach faced difficulties maintaining high-quality, personalized customer interaction while managing an increasing volume of inquiries, especially during peak travel seasons.

Challenge: Their primary obstacle was to scale their customer service without losing the personal touch that set them apart. They needed a solution that could handle routine inquiries and provide information in a manner that was consistent with their brand's friendly and adventurous spirit.

Solution: The agency decided to implement and customize ChatGPT to reflect its unique brand voice and cater to specific customer needs. The process, prompts, and outcomes involved are listed below.

Prompt: Adopt a conversational and enthusiastic tone when discussing travel destinations and itineraries.

ChatGPT provided a clear brand voice and guidelines that reflected an adventurous, friendly, and informative tone.

Prompt: Use our in-house travel offers and procedures [uploaded content] to offer detailed and personalized travel advice.

ChatGPT provided custom content with access to their extensive travel guides, blogs, and itineraries.

Prompt: Provide up-to-date availability and pricing for travel packages upon request.

ChatGPT was integrated with their booking system to offer real-time availability and pricing.

Prompt: Provide detailed answers to frequently asked questions about travel documentation and insurance.

Custom prompts were created to guide ChatGPT in handling common travel-related queries.

Prompt: Update responses based on the latest travel advisories and customer feedback trends.

> They established a system to regularly update ChatGPT based on customer feedback and new travel trends.

Results: After implementing the customized ChatGPT, this travel agency reaped the benefits of increased customer service efficiency, booking rates, and customer satisfaction.

- **Customer Service Efficiency:** There was a 53% reduction in response time for customer inquiries, with 24/7 availability.
- **Increased Booking Rates:** They saw a 23.6% increase in conversions from inquiries to bookings, attributed to timely and personalized responses.
- **Customer Satisfaction:** Post-implementation surveys showed a 37.2% increase in customer satisfaction regarding the quality and speed of responses.
- **Brand Consistency:** The travel agency reported that ChatGPT consistently adhered to its brand tone, enhancing its brand image.

Conclusion: This travel agency's experience illustrates the effectiveness of customizing ChatGPT to align with a business's specific needs and brand identity. By tailoring ChatGPT to their unique brand voice and integrating it with their existing systems, they enhanced customer interaction and improved response efficiency. They maintained their reputation for personalized service, even while handling a larger volume of inquiries. This case study exemplifies how businesses can leverage AI customization to elevate customer experience and streamline operations without sacrificing their brand's identity and unique personal touch.

ASK THE BOT: ADDITIONAL AI COMMANDS

To assist you in applying the knowledge you learned from this chapter on how to set up, navigate, and customize ChatGPT, open up and log into ChatGPT on your mobile device or web browser, and enter the following queries.

> I'm new to ChatGPT. Can you walk me through the initial setup process for integrating ChatGPT into the operations for [enter your business, website, or industry] and tell me what I need to get started?

> Once ChatGPT is set up for [enter your business, website, or industry], how do I navigate and interact with the interface effectively?

> How can I customize ChatGPT's responses to provide a more personalized experience for our customers for [enter your business, website, or industry] and align ChatGPT's responses with our business's tone and style?

> Can you explain the process of training ChatGPT to understand [enter the name of your industry] or specific customer needs of [enter the name of your company and website] better? What type of data or prompts should I use to train it effectively?

> Once ChatGPT is up and running, how do I ensure it continues to perform optimally and what are the ongoing monitoring and maintenance tasks I should be aware of to keep it effective?

These prompts will help any small business get started with ChatGPT, navigate its interface, and understand the customization process and ongoing management for the best results.

ADDITIONAL TRAINING

Since ChatGPT is continually evolving and each industry is unique, we have created extensive online training programs on our website, www.DynamicChatGPT.com, to further assist you in harnessing the power of AI for your small business.

CHAPTER 4
TIPS FOR MAXIMIZING CHATGPT'S POTENTIAL

WHEN WE WERE FIRST INTRODUCED to AI and ChatGPT, we weren't convinced it could help us in any of our businesses. We were beyond skeptical, viewing it as a feeble attempt to disguise a new vegetable on our plate by pouring cheese over it. We believed the reason was that very few "experts" we had spoken to had a grasp of the understanding of AI's vast capacities. But by having been on the cutting edge of other technologies in the past, we knew we needed to do our research and find experts who could explain this. We had one advantage over most people: we both have strong technical IT backgrounds where we could deep dive into this information without being overwhelmed. We realize that might not be the case with most of our clients, which is why we wrote this book in a format that anyone can understand.

We'll explore valuable tips and best practices for getting the most out of ChatGPT and maximizing AI's potential for your business. Whether using it for customer support, content generation, or any other purpose, following these guidelines will help you maximize ChatGPT's potential while avoiding common pitfalls and staying updated with AI advancements.

BEST PRACTICES FOR EFFECTIVE USAGE

Adopting certain best practices is essential to maximize the effectiveness of ChatGPT in your business. These guidelines will help ensure that ChatGPT operates efficiently, aligns with your business goals, and delivers a positive user experience.

1. Clearly Define Objectives

Begin by describing your specific goals and objectives for using ChatGPT. The more information you give AI, the better your results. Whether providing customer support, generating content, or automating tasks, clarity of purpose is paramount to your success.

2. Know That Data Quality Matters

Ensure that the data you use to train or fine-tune ChatGPT is of the highest quality and relevance to your specific use case. Clean and relevant data always leads to better performance, especially with AI. Hopefully, you've seen this with the queries you performed at the end of each chapter.

3. Monitor and Supervise

As you would any new process or technology, you need to continuously monitor ChatGPT's interactions, especially during the initial stages of deployment. This allows you to correct any inaccuracies and improve its responses over time.

4. Set Appropriate Boundaries

Establish clear boundaries for ChatGPT to ensure it stays within ethical and legal limits. Train it to avoid sensitive topics, misinformation, or inappropriate content. The more you use this tool, the more effective you and it will become.

5. Use Human Oversight

Incorporate human oversight into your AI processes. We refer to AI as an excellent soup starter. You must still add quality ingredients (information) for your output to be workable (or edible)! While ChatGPT can automate many tasks, human judgment is imperative, particularly for complex issues or sensitive situations.

6. Incorporate Feedback Loops

Implement mechanisms to gather feedback from users about their interactions with ChatGPT. Use this feedback to make iterative improvements to the AI model, enhancing its effectiveness and accuracy.

7. Provide Escalation Paths

While ChatGPT can handle many queries, some may require human intervention. Ensure there are clear escalation paths for complex issues or when users wish to speak with a human representative.

8. Educate Your Team

Make sure your team understands how ChatGPT works and its role within your business. This helps align its usage with your business processes and ensures smooth integration with your existing workflows.

9. Test and Validate Regularly

Periodically test the system to validate its performance and identify any issues. Regular testing helps in maintaining the efficiency and reliability of ChatGPT.

10. Prepare for Technological Changes

Stay informed about advancements in AI and be prepared to adapt your ChatGPT integration accordingly. Technological evolution can offer new opportunities to enhance the capabilities of your AI assistant.

By following these best practices, you can ensure that ChatGPT is an effective and integral part of your business strategy, enhancing operational efficiency and improving customer experience.

AVOIDING COMMON PITFALLS

To effectively leverage ChatGPT in your business, it's important to be aware of and avoid common pitfalls hindering its performance and utility. Recognizing and knowing how to address these potential issues will ensure that ChatGPT serves your business objectives more effectively.

1. Over-reliance on AI

Avoid over-reliance on AI. While ChatGPT can handle numerous tasks, it's not a one-size-fits-all solution and may not replace the need for human expertise in specific contexts.

As an example, a small healthcare clinic uses AI to manage incoming patient queries online. They set up ChatGPT to answer various questions, from appointment scheduling to medical advice. Over time, the clinic started relying too heavily on ChatGPT to handle not just administrative queries but also more complex medical inquiries. While ChatGPT efficiently manages routine questions, it lacks the expertise and nuanced understanding required for medical advice.

A situation arose where a patient used the ChatGPT service to inquire about symptoms they were experiencing. ChatGPT, operating on its pre-programmed knowledge, provides generic health advice, not accounting for the patient's specific medical history or the potential seriousness of the symptoms.

The patient, relying on this advice, delays seeking professional medical attention, leading to a worsening of their condition. This incident not only caused harm to the patient but also exposed the clinic to liability issues and damaged its reputation.

This example underscores the risk of over-relying on AI for tasks that require professional judgment, expertise, and a human touch, especially in critical fields like healthcare. It highlights the importance of maintaining a balance between leveraging AI for efficiency and ensuring expert human oversight for complex and sensitive matters.

2. Lack of Training

Insufficient training or neglecting to fine-tune ChatGPT for your specified needs can result in suboptimal performance. Invest time in training to achieve better results. If a team member doesn't know how to update ChatGPT's knowledge base, critical new product information might not be added, which leads to outdated or incorrect responses.

Imagine a local bookstore that decides to use ChatGPT to engage with customers online, answering inquiries about book availability, author events, and personalized reading recommendations. Enthusiastic about this technology, they quickly set up ChatGPT on their website without adequately training it on their specific inventory, events, and customer service style. As a result, ChatGPT starts to provide generic responses that don't accurately reflect the bookstore's unique collection or the local authors they promote. When customers ask for recommendations, AI fails to suggest books that are actually available in the store or upcoming local author events, creating misinformation and disappointment.

Customers start to feel that the AI service lacks the personal touch and expertise they expect, leading to frustration and decreased engagement with the bookstore. The staff, not trained on how to update and manage the AI's responses, struggle to rectify the situation.

This example highlights the importance of proper training and updating ChatGPT regularly to ensure it aligns with the business's specific services, products, and customer interaction style. It also underscores the need for staff training in managing and maintaining AI tools to maximize their effectiveness.

3. Ignoring Ethical Concerns

Ignoring ethical AI-related concerns, such as bias or privacy, can lead to reputational damage and legal issues. Address these concerns proactively in your AI implementations.

Imagine a business using ChatGPT for customer service without considering ethical guidelines or sensitive topics. A customer might ask ChatGPT about a product suitable for a religious event. If AI is not programmed with cultural and religious sensitivities, it might recommend inappropriate products or make culturally insensitive comments, which can offend the customer and others who read the interaction.

This kind of oversight leads to immediate customer dissatisfaction and can spark broader outrage and go viral if shared on social media and other platforms. This highlights the importance of instilling ethical considerations in AI systems to respect cultural, religious, and personal sensitivities, ensuring respectful and appropriate interactions.

4. Neglecting User Experience

Prioritizing user experience when implementing ChatGPT is central to AI's success in your organization. Poorly designed interfaces or confusing interactions can deter users and hinder the effectiveness of the AI. Customers might report that ChatGPT repeatedly misunderstands a particular query, but if this feedback is ignored, the issue will persist, leading to customer frustration.

Let's consider an online retail fashion store that employs ChatGPT to answer customer queries about products, sizing, and order processes. Initially, the store programmed ChatGPT to provide detailed, comprehensive answers, thinking this would be helpful to customers.

However, they didn't account for the user experience aspect, particularly the need for quick, concise answers in the context of an online shopping experience. Customers started to find the lengthy, complex responses from ChatGPT overwhelming and time-consuming, especially when they needed quick answers to simple questions like the availability of sizes or return policies.

This led to a frustrating shopping experience for many users, who preferred straightforward and easy-to-digest information. As a result, this company noticed a significant drop in customer satisfaction and an increase in abandoned shopping carts, as customers were turned off by the cumbersome interaction with the AI.

This example highlights the importance of considering user experience in ChatGPT interactions, ensuring responses are user-friendly, context-appropriate, and aligned with the customers' needs and preferences.

5. Stagnation

Since AI is continually evolving, failing to keep up with advancements can quickly leave businesses using outdated technology. Staying informed about the latest developments in AI is key to remaining competitive.

As an example, a travel agency integrated ChatGPT into its website a few years ago to assist customers with travel inquiries, bookings, and advice. Initially, the AI model provided great value, answering questions effectively and improving customer engagement.

However, over time, advancements in AI led to newer, more sophisticated models capable of more nuanced understanding and interaction, including updated information about travel restrictions, health advisories, and personalized travel recommendations based on evolving trends. Despite these advancements, the travel agency continued using the older version of ChatGPT without checking for updates.

As a result, the agency's AI began to fall behind in terms of capabilities. Customers started noticing that the ChatGPT responses were less relevant, especially when dealing with new travel concerns like pandemic-related restrictions or the latest eco-friendly travel options. Competitors who upgraded to newer AI models offered better, more informed interactions, leading customers to favor those agencies for their travel needs.

This stagnation not only caused the agency to lose its competitive edge but also reflected poorly on its commitment to leveraging cutting-edge technology for customer service, ultimately impacting its reputation and business success. This situation demonstrates the importance of staying updated with the latest AI advancements to ensure the technology remains effective, relevant, and competitive.

By being aware of all these pitfalls and actively working to avoid them, businesses can ensure that their implementation of ChatGPT is effective, efficient, and well-received by both customers and staff.

STAYING UPDATED WITH AI ADVANCEMENTS

To ensure you stay current with the rapid advancements in AI and maximize the potential of tools like ChatGPT, it's important to actively engage with various sources of knowledge and innovation in the field. At the time of print, specific suggestions include but are not limited to the activities listed below.

1. Follow AI News

Since AI is rapidly changing, following Artificial Intelligence news sources, academic journals, and industry publications will assist you in staying informed of the latest breakthroughs and trends.

- **Sources:** Reviewing AI-focused news platforms, blogs, and websites is helpful in staying abreast of technological advancements. Also, subscribing to newsletters from leading AI research institutions or tech companies is useful for business owners.
- **Academic Journals:** Stay abreast of the latest research by reading academic journals that publish AI and machine

learning papers, such as the *Journal of Artificial Intelligence Research* or *Neural Information Processing Systems*.

- **Industry Publications:** Following industry-specific research publications, such as Gartner, G2, Forrester, etc., that discuss AI applications in different sectors, providing insights into how AI is transforming various industries.

2. Attend Conferences and Webinars

Nothing beats networking at conferences and webinars. Attending local, national, and international AI conferences, webinars, and workshops to gain insights from experts and network with professionals in the field can prove helpful with current and future AI advancements.

- **Conferences:** Currently, the most popular AI conferences are NeurIPS, ICML (International Conference on Machine Learning), or industry-specific AI events. These conferences often feature talks by leading researchers and practitioners.
- **Webinars and Workshops:** Participate in webinars and online workshops. Many are free and can provide valuable insights into current trends and best practices. Search the internet for the most up-to-date listing in your area and online. We also offer training on our website, www. DynamicChatGPT.com.

3. Engage in Online Communities

Join online AI communities and forums where you can exchange ideas, ask questions, and learn from others in the AI community.

- **Forums and Groups:** AI-focused forums and online communities such as Stack Overflow, Reddit's machine learning community, or LinkedIn groups dedicated to AI and machine learning are beneficial to your business successfully integrating AI.
- **Q&A Sessions:** Participating in Q&A sessions and discussions within these communities to get answers to

specific questions and learn from real-world experiences of AI professionals is extremely helpful as you navigate the AI landscape.

4. Experiment and Innovate

Encourage experimentation and innovation with your team and organization. Again, since AI is a rapidly evolving field, being open to trying new approaches can lead to significant advancements in your company.

- **Internal Projects:** Encourage your team to undertake small-scale AI projects to experiment with new ideas. This hands-on experience can be invaluable.
- **Innovation Culture:** Foster a culture of innovation where employees feel comfortable proposing and trying out new AI-based solutions or improvements.

5. Collaborate

Network and connect with AI researchers, developers, and experts to explore new possibilities and stay at the forefront of AI technology.

- **Partnerships with Academia:** Forge collaborations with universities or research institutions. This can provide access to cutting-edge research and talented individuals in the AI field.
- **Industry Collaboration:** Network with other businesses and AI experts. Collaborations can lead to shared knowledge and potentially beneficial partnerships for implementing advanced AI solutions.

Maximizing ChatGPT's potential requires a combination of utilizing practical usage, avoiding common pitfalls, and staying updated with AI advancements. Following these tips and best practices ensures that ChatGPT remains a valuable asset in your business operations. Throughout the remainder of this book, you'll have a comprehensive understanding of how AI, and specifically

ChatGPT, can uniquely benefit your small business across various domains.

CASE STUDY: IMPLEMENTING BEST PRACTICES FOR CHATGPT

A small boutique art supply store in the southern U.S. prides itself on providing a curated selection of high-quality art materials and personalized customer service. Despite a loyal customer base, they struggled to efficiently manage online customer inquiries and provide tailored product recommendations, especially during peak holiday seasons.

Challenge: The store needed a solution to manage growing online customer inquiries without compromising the customizable services that distinguished them from their competition. Using their small team, they sought an efficient way to handle queries while maintaining their standard of service.

Solution: They integrated ChatGPT into their website and social media platforms, adhering to the best practices outlined in their recent AI strategy revamp. The prompts used and outcomes are below.

Prompt: Provide quick and personalized product recommendations based on customer preferences in art supplies to reduce customer response time by 50% and increase personalized product recommendations [inputed data].

Outcome: Improved response time for customer inquiries, offered personalized product recommendations, and automated routine queries.

Prompt: Use the following product catalog to suggest the best art supplies for watercolor painting: [Inputed product catalog, customer FAQs, and specific art-related queries].

Outcome: ChatGPT provided detailed product descriptions, usage guides, and customer reviews.

Prompt: Provide a schedule and procedures to regularly review and update responses to maintain accuracy, brand, and tone.

> Outcome: The team implemented weekly audits of ChatGPT conversations and reviewed ChatGPT's interactions, making adjustments to ensure responses were accurate and aligned with their brand voice.

Prompt: If a query is unrelated to art supplies, guide the customer to contact a human representative to set appropriate boundaries.

> Outcome: They configured ChatGPT to avoid sensitive topics and maintain a focus on art-related queries and created guidelines for ChatGPT to redirect unrelated queries to human representatives.

Prompt: For complex art project inquiries, suggest connecting with our art consultants.

> Outcome: The store staff supervised ChatGPT, particularly for complex art-related inquiries or when personal shopper experience was requested. Art consultants also reviewed ChatGPT suggestions for custom art supply kits.

Prompt: At the end of each interaction, ask customers to rate their experience.

> Outcome: They implemented a simple rating system feedback mechanism at the end of each chat interaction for customers to rate their ChatGPT interaction, using this data to refine the AI's performance.

Prompt: Offer customers the option to speak with a store representative for more complex needs.

Outcome: They created an escalation path where complex inquiries were routed to human staff, with clear indications for customers on how to request human assistance.

Prompt: Assist our staff in integrating AI recommendations into personalized service.

Outcome: Staff were trained on how ChatGPT functions and how to utilize resources in conjunction with their roles. They created workshops on integrating ChatGPT recommendations for more personalized customer service.

Prompt: Conduct monthly performance tests using simulated customer inquiries.

Outcome: They set up monthly tests to assess ChatGPT's performance, make necessary adjustments, and implement testing procedures for simulating customer inquiries to evaluate response accuracy and relevance.

Prompt: Stay updated on the latest AI features and integrate them into our service model.

Outcome: They stayed updated on AI advancements, which made them ready to integrate new features to enhance ChatGPT's capabilities. They also subscribed to AI newsletters and initiated regular consultations with an AI expert.

Results: After six months of using ChatGPT, this art supply store observed enhanced efficiency, increased sales, and improved customer satisfaction.

- **Enhanced Efficiency:** The art store experienced a 61.4% reduction in response time for online inquiries.
- **Increased Sales:** Personalized product recommendations by ChatGPT increased sales by 27%.

- **Customer Satisfaction:** Improved customer satisfaction scores by 47.2%, particularly regarding the speed and relevance of responses.
- **Team Productivity:** Staff were able to focus more on in-store customers and complex online queries, improving overall service quality.

Conclusion: After implementing ChatGPT, this art supply store significantly improved its online customer interaction, sales, and operational efficiency. This case study exemplifies how small businesses, even in niche markets, can effectively leverage AI tools like ChatGPT by adhering to a structured set of best practices tailored to their unique business needs and goals. This success story underscores the potential of AI to enhance customer experience and business operations when strategically and thoughtfully implemented.

ASK THE BOT: ADDITIONAL AI COMMANDS

To assist you in applying the knowledge you learned from this chapter, here are five queries for small businesses seeking tips to maximize ChatGPT's potential with best practices, avoid common pitfalls, and stay updated with AI advancements. Open up and log into ChatGPT on your mobile device or computer and enter the subsequent queries.

> What are the best practices for using ChatGPT effectively within [enter your business, website, or industry]? How can we ensure that we leverage its capabilities to the fullest?

> We want to make the most of ChatGPT but also avoid common mistakes. What common pitfalls do businesses in [enter your industry] face when implementing AI like ChatGPT, and how can we steer clear of them?

> Since the AI landscape is constantly evolving, how can we stay updated with the latest advancements in AI, including updates and improvements related to ChatGPT, to maintain a competitive edge for [enter your business, website, or industry]?

> To align ChatGPT with our unique business needs, what strategies and approaches should we consider to effectively customize its responses and functionality for [enter your business, website, or industry]?

> Since the AI landscape is constantly evolving, how can we stay updated with the latest advancements in AI, including updates and improvements related to ChatGPT, to maintain a competitive edge for [enter your business, website, or industry]?

> As [enter your business, website, or industry] grows, how can we scale our usage of ChatGPT to accommodate increased demand and ensure a consistent, high-quality AI-driven customer experience?

Enter these queries into ChatGPT to see how they can help your small business gain valuable insights into making the most of this new tech while avoiding common pitfalls and staying informed about AI advancements in your industry.

ADDITIONAL TRAINING

Since ChatGPT is continually evolving and each industry is unique, we have created extensive online training programs on our website, www.DynamicChatGPT.com, to further assist you in harnessing the power of AI for your small business.

CHAPTER 5
USING CHATGPT FOR CUSTOMER SUPPORT

WHEN RON STARTED his first business over thirty-five years ago, he was the manager, the tech support, the sales trainer, the janitor, and the Director of Customer Service. That's a lot of hats! We are sure many small business owners can relate. At that time (back in the late 1980s), Ron and his partner owned a cutting-edge telecom/voicemail/cellular corporation. He spent much of his day on the phone with clients, providing both customer and technical support. While these activities create a positive image for a company, they don't generate immediate income. Today, most small businesses run the same way, but thankfully, with AI, they don't have to.

In this chapter, we'll explore how you can leverage ChatGPT to enhance your customer support operations. From building a ChatGPT-powered chatbot to efficiently handling inquiries and providing around-the-clock support, we'll cover the integral aspects of using ChatGPT for superior customer service.

BUILDING A CHATGPT-POWERED CHATBOT

A chatbot is an AI-powered tool that can interact with customers, answer questions, and provide assistance in a conversational manner. ChatGPT can be the foundation for building an effective chatbot for your small business. Since these platforms are continually being updated, check our website, www.DynamicChatGPT.com, for the latest technology recommendations. Meanwhile, here's how to get started with your new chatbot.

DEFINE YOUR CHATBOT OBJECTIVES

Stating clear objectives for your chatbot is the first critical step in ensuring it effectively aligns with your business needs. Determine what types of customer inquiries your bot will handle and the level of support it should provide.

1. Identify the Primary Function of the Chatbot

- **Customer Service:** If the main goal is to handle customer service inquiries, establish the range of questions the chatbot should be capable of addressing, like product information, order status, or troubleshooting.
- **Sales Support:** If the chatbot is intended to aid in sales, identify how it should assist in the buying process, perhaps by providing product recommendations, answering product FAQs, or guiding customers through the checkout process.

2. Determine the Scope of Queries

- **General Inquiries:** Designate if the chatbot will handle general questions about your business, like hours of operation, location information, or services offered.
- **Specialized Support:** If your business requires specialized knowledge, such as technical support for electronics or detailed information about financial services, the chatbot might need to be equipped with more specific information.

3. Level of Interaction

- **Basic Interaction:** For simple tasks, the chatbot might only need to provide predefined answers to common questions.

- **Advanced Interaction:** For more complex engagements, like providing personalized recommendations or handling nuanced customer service issues, the chatbot will need advanced capabilities, possibly integrating ChatGPT with similar AI models.

4. Integration with Business Processes

- **Order Processing:** If the chatbot will be part of an e-commerce platform, stipulate how it will integrate with order processing systems, like checking order status or updating delivery information.
- **Appointment Scheduling:** For service-oriented businesses, the chatbot might need to integrate with scheduling systems to book, reschedule, or cancel appointments.

5. User Experience Goals

- **Ease of Use:** The chatbot should be easy to interact with, providing clear and concise information that aligns with your corporate voice, identity, and message.
- **Engagement:** Define how the chatbot should engage users, perhaps through a friendly tone or by providing interactive elements like quick reply options.

6. Performance Metrics

- **Response Accuracy:** Set business goals for how accurately the chatbot should respond to inquiries.
- **Resolution Rate:** Determine the expected rate of successfully resolved queries without human intervention.

7. Continuous Improvement

- **Feedback Loop:** Establish a mechanism for collecting user feedback to continually improve the chatbot's accuracy.

By clearly defining these objectives and expectations, you can ensure that the development and implementation of your chatbot are focused and aligned with your business goals, ultimately leading to a more successful deployment and a better experience for your customers.

CHATBOT DEVELOPMENT PLATFORMS

Choosing the right chatbot development program is paramount for effectively integrating and leveraging ChatGPT in your business. These applications often provide user-friendly interfaces, making the design and deployment of chatbots more accessible. At the time of print, here are some current examples and unique features of chatbot development platforms that can integrate ChatGPT.

1. Dialogflow (by Google): Offers natural language understanding and is excellent for creating conversational interfaces. It can be integrated with websites, mobile apps, and popular messaging platforms like Facebook Messenger. You can incorporate ChatGPT into Dialogflow to enhance its conversational capabilities, making it more dynamic and context-aware.

Imagine a small business that specializes in artisan coffee and homemade pastries. They want to enhance their customer service and online presence. The owner integrated Dialogflow, powered by Google, with their website and Facebook Messenger.

Dialogflow offers natural language understanding, creating a conversational interface that feels natural and engaging. To enhance this experience further, they decided to incorporate ChatGPT into Dialogflow. This integration makes their digital assistant more dynamic and context-aware, capable of handling various customer inquiries about menu items, opening hours, and special offers. It can even take orders or book reservations. The visual reflects a modern, inviting café atmosphere with a digital screen showing a friendly chat interface, demonstrating the seamless integration of Dialogflow and ChatGPT in a small business setting. This scenario showcases how a small business can leverage advanced conversational AI to improve customer engagement and operational efficiency using Dialogflow and ChatGPT.

2. Microsoft Bot Framework: Provides tools for building, testing, deploying, and managing intelligent bots. This bot also allows integration with various channels like Skype, Slack, and Microsoft Teams. You can also establish ChatGPT into MS Bot Framework to create more sophisticated bots for complex conversations and tasks.

A small tech startup that focuses on delivering cutting-edge IT services desires to streamline its customer support and internal communication. To achieve this, they utilized the Microsoft Bot Framework, known for its robust tools in building, testing, deploying, and managing intelligent bots.

The startup integrates the bot with various channels, including Skype, Slack, and Microsoft Teams, to ensure seamless communication across all platforms. To enhance the bot's capabilities, they also incorporate ChatGPT. This integration allows the bot to engage in more sophisticated conversations and perform complex tasks, such as troubleshooting technical issues, scheduling meetings, or providing detailed product information. The visual depicts a lively startup office environment, with employees interacting smoothly with a sophisticated bot interface on various devices. The focus would be on the seamless integration of Microsoft Bot Framework and ChatGPT, highlighting how it revolutionizes communication and efficiency in a small business setting.

This instance illustrates the transformational impact of combining Microsoft Bot Framework and ChatGPT in improving both customer and internal communications for a small business.

3. IBM Watson Assistant: Known for its powerful AI and natural language processing capabilities. IBM Watson Assistant allows for the creation of advanced chatbots that can understand nuances in language. By merging ChatGPT into Watson Assistant bots, customer interactions can be further enhanced to deliver more personalized and contextually relevant interactions.

A renowned travel agency seeks to revolutionize its customer service. They chose IBM Watson Assistant for its exceptional AI and natural language processing capabilities, enabling them to create advanced chatbots that understand the nuances of language. The chatbots are designed to assist customers in planning their trips, offering personalized suggestions for destinations, accommodations, and activities based on the customer's preferences and past travel history.

To enhance these interactions further, they integrate ChatGPT into their Watson Assistant bots. This integration allows the chatbots to conduct more in-depth and contextually relevant conversations with customers. For example, a customer inquiring about a beach holiday in Thailand would receive tailored recommendations for beach resorts, local cuisine, and must-visit attractions, along with real-time weather updates and travel advisories.

The user interface depicts a user engaging with their website, where the chatbot, powered by IBM Watson Assistant and ChatGPT, provides detailed and personalized travel advice. The screen shows a friendly and intelligent conversation flow, highlighting the chatbot's ability to understand complex queries and offer detailed, customized responses, thereby enhancing the overall customer experience in the travel planning process.

This example demonstrates how the synergy between IBM Watson Assistant and ChatGPT can create a highly personalized and efficient customer service experience in the travel industry.

4. Chatfuel: Primarily focused on Facebook Messenger bots. It offers a straightforward interface for creating bots without the need for coding. ChatGPT works with Chatfuel to provide more advanced conversational abilities, making the bots better at handling a wide range of customer queries.

Visualize a flourishing online boutique that specializes in contemporary fashion. They're eager to engage more interactively with their customers on social media. To achieve this, they opt for Chatfuel, renowned for its ease of creating Facebook Messenger bots without requiring coding skills. This choice aligns perfectly with their goal to have an accessible and responsive digital presence.

They enhance their Chatfuel bot by integrating it with ChatGPT. This integration dramatically improves the bot's conversational abilities, making it adept at handling a diverse range of customer queries. Whether a customer is asking for fashion advice, inquiring about the latest trends, checking order status, or seeking help with returns, the bot is equipped to provide prompt and accurate responses to all inquiries.

The interface showcases a customer interacting with their Facebook Messenger bot. The conversation on the screen reflects the bot's advanced capabilities, thanks to ChatGPT, displaying how it adeptly navigates complex queries and offers personalized fashion advice. This interaction highlights the bot's enhanced functionality in providing a seamless and engaging customer service experience on social media.

5. ManyChat: A popular platform for creating Facebook Messenger and SMS chatbots. Offers easy-to-use tools for bot design and deployment. Adding ChatGPT to ManyChat bots can significantly improve their responsiveness and ability to engage in more meaningful conversations.

Picture a thriving fitness center renowned for its personalized training programs and health-centric community. They decided to enhance their member engagement and support by leveraging ManyChat, a popular platform known for its user-friendly tools for creating chatbots for Facebook Messenger and SMS. They use ManyChat to design a bot that assists members with workout queries, class schedules, and nutrition tips.

To take their service to the next level, they integrate ChatGPT with their ManyChat bot. This combination significantly boosts the bot's responsiveness and elevates its ability to engage in meaningful conversations. Members can now receive more personalized fitness advice, timely reminders for their workout sessions, and motivational messages directly through Facebook Messenger or SMS.

The interface captures a situation where a gym member is interacting with their chatbot on their smartphone. The screen indicates a dynamic conversation where the bot, empowered by ChatGPT, provides tailored workout suggestions, answers complex health-related questions, and offers motivational support. This interaction exemplifies how the integration of ChatGPT with ManyChat can transform member communication and support in a fitness center setting.

6. Rasa: An open-source platform ideal for developers looking to build more customized and sophisticated chatbots. Including ChatGPT into Rasa can lead to highly sophisticated bots capable of understanding complex queries and maintaining context over longer conversations.

Imagine a dynamic startup that specializes in innovative software development. They are committed to providing top-notch customer support and streamlining their project management process. To achieve this, they turn to Rasa, an open-source platform ideal for developers looking to build customized and sophisticated chatbots. They use Rasa to develop a bot tailored to their specific needs, capable of handling technical support, project updates, and client queries.

To further enhance the bot's capabilities, they integrated ChatGPT into their Rasa bot. This integration leads to the creation of a highly sophisticated bot that not only understands complex technical queries but also maintains context over more extended conversations. The bot can assist clients with detailed software inquiries, provide updates on project timelines, and offer solutions to technical challenges.

The illustration portrays a client engaging with their chatbot on a sleek, modern interface. The interaction illustrates the bot's advanced conversational skills, seamlessly handling a complex software development query, remembering past interactions, and providing contextually relevant responses. This interaction showcases how the integration of ChatGPT with Rasa elevates the customer support experience in a high-tech software development environment.

7. Botpress: An open-source bot builder that offers a visual conversation builder known for its extensibility and customizability. By incorporating ChatGPT, Botpress bots can be made more intelligent, offering better and more natural user interactions.

A local bookstore is known for its cozy atmosphere and personalized service. Seeking to enhance its online customer interaction, this store uses Botpress, an open-source bot builder known for its visual conversation builder and extensive customizability. They designed a Botpress bot to assist online visitors with book recommendations, store hours, event information, and online order inquiries.

To take their digital customer service to the next level, ChatGPT is integrated with the Botpress bot. This integration significantly enhances the bot's intelligence, offering more natural and engaging user interactions. Now, customers can receive personalized book recommendations based on their reading preferences, detailed synopses, and author backgrounds, and even participate in book club discussions, all through the bot.

The user interface shows a customer interacting with the website, where the Botpress bot, powered by ChatGPT, is engaging in a detailed conversation about book recommendations. The chat window displays the bot's ability to understand and respond to complex queries about genres, authors, and specific book titles, demonstrating how the integration with ChatGPT has made the bot more intelligent and user-friendly.

This illustrates how integrating Botpress with ChatGPT can transform the customer service experience in a retail environment, particularly for a local bookstore aiming to offer personalized online engagement.

Each of these platforms can significantly and uniquely enhance the chatbot's capabilities, providing users with a more sophisticated and helpful interaction experience. Research each one for your organization and desired use regarding the level of customization, platforms where the chatbot will be deployed (like websites, social media, or internal communication tools), and intended goals and metrics for the chatbot.

INTEGRATING THE CHATBOT WITH CHATGPT

The next step is integrating a chatbot with ChatGPT, which involves configuring the chatbot to interact with the AI model effectively. This process ensures that user queries are processed by ChatGPT, and the responses are relayed back through the chatbot interface seamlessly. Below is a more detailed approach to this integration. If you have questions on the specific steps, don't hesitate to ask ChatGPT for more detailed instructions.

1. **Establish API Connectivity:** Utilize the provided APIs (Application Programming Interfaces) to connect your chatbot with ChatGPT. This involves setting up API calls from your chatbot to ChatGPT and handling responses from ChatGPT.
2. **Configure Query Handling:** Implement logic within your chatbot to correctly format and send user queries to ChatGPT. Confirm that the chatbot can interpret user inputs in an understandable way for ChatGPT and your team.
3. **Handling Responses:** Once ChatGPT processes the query and generates a response, your chatbot should be configured to receive this response and present it to the user in a clear, user-friendly format.
4. **Context Management:** Ensure that the chatbot maintains the context of the conversation. This may involve storing dialogue and their stats so that ChatGPT can provide contextually relevant responses.
5. **Fallback Mechanisms:** In cases where ChatGPT is unable to provide a satisfactory answer, have a fallback mechanism in place, such as escalating to a human agent.
6. **Performance Optimization:** Regularly monitor the interaction between your chatbot and ChatGPT. Use this data to optimize your query handling and response generation for better performance and user satisfaction.
7. **Data Privacy and Security:** Ensure the integration adheres to privacy laws and protocols, especially when handling user information. As always, treat your chatbot as you would any third-party software agent with sensitive data.
8. **User Feedback Integration:** Implement a system to collect user feedback on the chatbot's performance, particularly the responses generated by ChatGPT. Use this feedback for continuous improvement.

By following these steps, you can create a robust integration between your chatbot and ChatGPT, enabling a seamless and effective user experience. In doing so, you can significantly enhance the capabilities of your chatbot, making it more intelligent, responsive, and useful to your customers.

TRAINING, FINE-TUNING, AND TESTING

Training and fine-tuning your chatbot in conjunction with ChatGPT is vital for ensuring its effectiveness and relevance in real-world situations. Before going live, thoroughly test the integration in various scenarios and examples to ensure that the chatbot and ChatGPT work together smoothly and efficiently. Also, test the chatbot to ensure it can handle common customer inquiries effectively and provide accurate responses. Below is an overview of the necessary steps.

1. **Data Collection**

- **Gather Relevant Data:** Compile a comprehensive data set with typical user queries, responses, and interaction patterns relevant to your business or application.
- **Previous Interactions:** Analyze logs of prior customer interactions to identify common queries, issues, and effective responses.

2. **Initial Training**

- **Base Training:** Use the collected data to train your chatbot, ensuring it has a foundational understanding of the types of queries it will encounter.
- **ChatGPT Queries:** Train the chatbot to effectively use ChatGPT for processing and responding to queries, ensuring the responses are relevant and accurate.

3. **Scenario-Based Testing**

- **Create Test Scenarios:** Develop a range of scenarios the chatbot might encounter, including common, complex, and edge-case scenarios.

- **Test Integration:** Run these scenarios to test how well the chatbot and ChatGPT work together, focusing on response accuracy, context handling, and user satisfaction.

4. Fine-Tuning

- **Identify Weaknesses:** Analyze the test results to identify areas where the chatbot struggles, such as certain types of queries or maintaining context over extended interactions.
- **Adjust and Optimize:** Modify the chatbot's configuration, training data, or how it interacts with ChatGPT to address these weaknesses.

5. Feedback Loop

- **Implement Feedback Mechanisms:** Set up ways to collect feedback from users during the testing phase and after going live.
- **Incorporate Feedback:** Continuously refine the chatbot based on this feedback, focusing on improving areas that users find lacking or confusing.

6. Continuous Learning

- **Update Training Data:** Regularly update the training data set with new queries, responses, and interaction patterns as they emerge.
- **Re-train Regularly:** Periodically re-train the chatbot to ensure it remains up-to-date with the latest information and interaction trends.

7. Monitoring and Evaluation

- **Performance Metrics:** Establish goals to evaluate the chatbot's performance, such as response accuracy, user satisfaction, and resolution rates.

- **Ongoing Assessment**: Continuously assess the chatbot against these metrics and make adjustments as needed.

8. User Experience Focus

- **User-Centric Design:** Ensure that the fine-tuning process always keeps the end user's experience in mind, aiming to make interactions as natural and helpful as possible.

By thoroughly training and fine-tuning your chatbot, especially in how it utilizes ChatGPT, you can significantly enhance its effectiveness and ensure that it provides a high level of service to your users. This ongoing process requires regular updates and adjustments to keep pace with changes in user behavior and expectations.

HANDLING FREQUENTLY ASKED QUESTIONS

One of the primary uses of ChatGPT in customer support is addressing frequently asked questions (FAQs). Here's how ChatGPT can assist small businesses in this regard.

- **FAQ Integration:** Upload your list of frequently asked questions and their answers to your chatbot. ChatGPT can reference this data to respond to common inquiries.
- **Instant Responses:** ChatGPT can provide instant responses to standard queries, freeing up human agents to focus on more complex customer issues.
- **Consistency**: Using ChatGPT to handle FAQs ensures consistent and standardized responses to customer queries, regardless of the time of day.

MANAGING CUSTOMER INQUIRIES EFFICIENTLY

Efficiency is vital in managing customer inquiries. ChatGPT can streamline the process and improve response times.

- **Quick and Accurate Responses:** ChatGPT can swiftly analyze and generate responses to customer inquiries, ensuring fast and accurate answers to their questions.
- **Handling Multiple Inquiries:** ChatGPT can handle multiple customer inquiries simultaneously, making it a valuable resource during peak hours.
- **Advanced Routing:** Implement an advanced routing system that directs inquiries to either human agents or a ChatGPT-integrated bot based on complexity. This ensures that human support agents address complex issues while the chatbot handles routine questions.
- **Integration with Existing Systems:** Meld ChatGPT-powered chatbots with your existing customer support systems, such as ticketing and CRM platforms, for a seamless customer experience.

PROVIDING 24/7 SUPPORT

One of the significant advantages of using ChatGPT for customer support is its ability to provide 24/7 assistance. This continuous availability can greatly enhance customer satisfaction and retention.

- **Around-the-Clock Support:** ChatGPT does not require rest or breaks, allowing your business to provide 24/7 customer support in different time zones.
- **After-Hours Assistance:** Customers who encounter issues or have questions outside regular business hours can still receive assistance, improving the overall customer experience.
- **Increased Customer Loyalty:** The availability of 24/7 support demonstrates your commitment to customer satisfaction, which can lead to increased loyalty and repeat business.
- **Handling Spikes in Inquiries:** During product launches, promotions, or other events that result in a surge of

inquiries, ChatGPT can ensure that customers receive timely responses, preventing frustration and lost opportunities.

Incorporating ChatGPT into your customer support strategy can lead to more efficient operations, improved customer satisfaction, and a competitive edge in the market. As we proceed through this book, you'll discover additional ways ChatGPT can benefit your small business, from content generation to data analysis and beyond.

CASE STUDY: ENHANCING CUSTOMER SERVICE WITH A CHATGPT-POWERED CHATBOT

A popular specialty plant store in the Midwest is renowned for its diverse houseplants and expert gardening advice. Despite its popularity, the store faced problems efficiently managing customer inquiries, particularly outside business hours and during peak gardening seasons.

Challenge: With a small team and a growing online presence, they needed a way to provide timely and informative responses to customer queries around the clock. To enhance their customer service without compromising the personalized advice to deliver the services that their customers valued, they utilized ChatGPT.

Solution: To address these goals, they developed a ChatGPT-powered chatbot for their website. The prompts used and outcomes are below.

> Prompt: Provide detailed care instructions for common houseplants and offer product recommendations based on our specific customer needs [inputed data].

> Outcome: The primary use cases were identified as answering common plant care questions, providing recommendations, and assisting with online order inquiries.

> Prompt: Explain the differences in watering needs between succulents and tropical plants.

Outcome: The chatbot was trained with specific plant care information, store policies, and FAQs to ensure accurate and relevant responses.

Prompt: Guide customers through our website to find products and information on plant care.

Outcome: The chatbot was integrated into their website with a user-friendly interface, making it easily accessible to online visitors.

Prompt: Conduct a series of tests simulating common customer queries about plant care and online orders.

Outcome: Before going live, the chatbot underwent extensive testing to ensure it could handle various inquiries effectively. Customer feedback was used for continuous improvements.

Prompt: Inform customers about the chatbot and encourage its use for quick plant care questions.

Outcome: The store actively promoted the chatbot through social media and in-store signage, encouraging customers to use it for quick queries.

Prompt: Regularly update the chatbot with new plant care advice and store information.

Outcome: The team regularly reviewed chatbot interactions to update its knowledge base and improve response accuracy.

Results: After three months of implementing the ChatGPT-powered chatbot, this organization experienced notable benchmarks.

- **Increased Customer Engagement:** Customer interactions increased by 69.1% during outside business hours.

- **Improved Response Time:** The average response time for online queries decreased from eight hours to immediate, 24/7 responses.
- **Enhanced Customer Satisfaction:** Customer feedback indicated an 88.1% satisfaction rate with the information and assistance provided by the chatbot.
- **Reduced Workload:** The staff reported a 39% decrease in routine query handling, allowing more focus on in-store customer service and other tasks.

Conclusion: By successfully leveraging a ChatGPT-powered chatbot to enhance its customer service, this specialty plant provider demonstrates how small businesses can use AI tools to improve efficiency and customer engagement. The chatbot provided immediate assistance to online customers and significantly reduced the staff's workload, enabling them to concentrate on other important aspects of the business. This case study highlights the potential of AI-powered chatbots in elevating the customer service experience, especially for companies with a specific niche and expertise.

ASK THE BOT: ADDITIONAL AI COMMANDS

To assist you in applying how to use ChatGPT for handling your business's FAQs, providing 24/7 customer support, and building a chatbot, open up and log into ChatGPT on your mobile device or computer and enter the following queries.

> How can I start building a chatbot using ChatGPT for [enter your business, website, or industry] to handle frequently asked questions and provide 24/7 customer support?

> What are the best practices to ensure ChatGPT provides helpful and relevant answers for training my chatbot to accurately understand and respond to customer inquiries for [enter your business, website, or industry]?

How do I customize the chatbot to reflect our brand's tone, style, and responses to ensure they align with [enter your business, website, or industry]'s messaging and values?

Once the chatbot is ready, how can I integrate it into [enter your business, website, or industry] or other communication channels, such as social media or messaging apps, to provide seamless customer support?

What metrics should I track to evaluate the performance of [enter your business, website, or industry]'s chatbot in handling customer inquiries? How can I continuously improve its effectiveness and user satisfaction?

These prompts will help your business understand how to build and utilize a ChatGPT-powered chatbot to handle frequently asked questions and offer around-the-clock customer support with your company's brand, style, and message.

ADDITIONAL TRAINING

Since ChatGPT is continually evolving and each industry is unique, we have created extensive online training programs on our website, www.DynamicChatGPT.com, to further assist you in harnessing the power of AI for your small business.

CHAPTER 6
AI CONTENT CREATION AND MARKETING

WE HAVE all sat in marketing meetings that have seemingly lasted an eternity. Several years ago, Ron was stuck in one that lasted two days. He and his team tried to develop a marketing campaign for a new territory when all their competitor's attempts had inexplicably failed. They struggled to create fresh ideas for this market. Perhaps you can relate?

On the other hand, Kim is a marketing guru. She can look at any business, analyze its mission statement and current marketing campaigns, and tell you what and where the disconnect is and how the company will succeed or fail. She truly has a gift and has used it to teach hundreds of university students and consult with businesses over the years. Many of us do not have this aptitude, but we have AI to assist us as small business owners.

As we all know, creating effective marketing for your business is a necessity on an ongoing basis. Today, you must utilize traditional as well as social media marketing. But what if that's not your strong suit? Or, what if you don't have enough staff experience to brainstorm a campaign to effectively increase revenues year over year?

In this chapter, we'll explore how ChatGPT can be a valuable asset in your content creation and marketing efforts. AI can significantly enhance your marketing strategies by generating engaging blog posts, crafting persuasive marketing copy, managing social media, and assisting with email marketing.

GENERATING ENGAGING BLOG POSTS AND ARTICLES

Blogging and content marketing are foundational in attracting and engaging your target audience. How often have you stared at your computer screen or phone, trying to generate a powerful story to propel your brand and product offerings? ChatGPT can assist in generating high-quality blog posts and articles that resonate with your readers.

- **Topic Ideation:** Use ChatGPT to brainstorm and generate ideas for blog topics based on trending industry keywords, customer interests, or emerging trends.
- **Content Outlines:** ChatGPT can create detailed outlines for your business industry articles, providing structure and clarity for your writing process.
- **Content Creation:** Automate the initial draft of your blog posts with ChatGPT. It can generate informative and engaging content to kickstart your writing process.
- **Editing Assistance:** After generating content, use ChatGPT to proofread and edit articles for grammar, style, and coherence.
- **Content Repurposing:** Reuse AI-generated content and scripts for various formats, such as infographics, videos, or podcasts, to expand your content marketing efforts.

CRAFTING PERSUASIVE MARKETING COPY

Creating compelling marketing copy is an art. Well-written copy can drive conversions and boost sales. It is vital for engaging potential customers and driving revenue. ChatGPT, with its advanced language processing capabilities, can be an invaluable tool in creating effective and compelling marketing materials. Here's an overview of how ChatGPT can assist in this process.

Analyzing Target Audience: Analyzing the target audience for small businesses involves understanding who the potential customers are, what they need or want, and how they make decisions. This process is foundational for small businesses to effectively tailor their products, services, and marketing efforts to meet their audience's specific needs and preferences.

- **Demographic Insights:** ChatGPT can analyze customer data to understand your target audience's demographic and psychographic profiles, helping tailor your marketing message to your preferences and interests.
- **Language Style:** It can adapt the style and tone of the copy to match the language that resonates best with your audience, whether it's formal, casual, or industry-specific jargon.

Creating Diverse Content Formats: Creating diverse content formats is an essential strategy for small businesses to engage their audience through various types of media and platforms. This approach caters to different content consumption preferences and helps reach a broader audience.

- **Different Channels:** ChatGPT can help generate a wide variety of brand messages suited for various platforms, keeping in mind the unique format and audience.
- **Ad Copy:** Use ChatGPT to generate advertising text for online advertising campaigns, including Google ads, social media ads, and more

Developing Unique Selling Propositions (USPs): This strategy for small businesses is one of the most powerful to differentiate themselves from their competitors, especially in a competitive market. A USP is a distinct feature or benefit that makes a business stand out from its competitors, making it the primary reason customers should choose it over others.

- **Highlighting Key Features:** ChatGPT can help articulate the unique benefits and features of your products or services, crafting your messages to highlight why your offering stands out from competitors.
- **Product Descriptions:** Craft persuasive product descriptions highlighting your offerings' unique features and benefits to entice potential customers.

Crafting Calls-to-Action (CTAs): This is an essential aspect of marketing and communication for small businesses. A CTA is a prompt or instruction to the audience to provoke an immediate response, usually involving an action like "buy now," "learn more," or "sign up." CTAs guide potential customers toward the next step in purchasing or engaging further with the business.

- **Effective CTAs:** ChatGPT can create compelling calls-to-action, which is fundamental for converting readers into leads or customers. It can suggest various CTA formats and messages to encourage user engagement from your target markets.

A/B Testing of Marketing Messages: This method compares two versions of a marketing message to determine which one performs better. This approach is critical for small businesses as it helps in making data-driven decisions about their marketing strategies.

- **Variation Generation:** Generate multiple versions of a single piece of copy, allowing you to test which variations are most effective in terms of customer engagement and conversion rates.
- **Testing Ideas:** Collaborate with ChatGPT to brainstorm A/B marketing messages, helping you refine your strategies for better results.

Integration with Marketing Strategies: This refers to the cohesive and strategic alignment of various marketing activities and channels to ensure a unified, consistent, and practical approach to promoting a brand, product, or service. This concept is vital for businesses of all sizes, including small businesses, as it helps maximize the impact of marketing efforts.

- **Campaign Alignment:** ChatGPT can ensure that your marketing messages and campaigns align with the overall marketing strategy and campaign themes, maintaining consistency across all content.
- **Landing Page Content:** Create captivating landing page content encouraging visitors to act, such as signing up for newsletters or purchasing.

Enhancing Storytelling: We have all heard the old adage, "Facts tell, but stories sell." This strategy is essential for small businesses and involves crafting and sharing compelling narratives that connect the company, its products, or services with its audience in a meaningful way. Effective storytelling can evoke emotions, build brand loyalty, and differentiate your offerings in a crowded market.

- **Narrative Development:** ChatGPT can assist in weaving compelling narratives around your brand or products, a technique proven to enhance customer engagement and brand recall.

Localization of Content: Adapting content to meet a specific target market or geographic area's cultural, linguistic, and contextual needs is extremely helpful for any small business. This goes beyond mere translation and involves customizing the business's messaging and materials to resonate with local audiences. Effective localization can significantly enhance the relevance and appeal of a business's offerings in different regions.

- **Adapting to Local Markets:** ChatGPT can help localize the marketing copy for businesses operating in multiple regions, ensuring it is culturally relevant and resonates with local audiences.

Regular Updates and Fresh Content: This is crucial for small businesses to maintain relevance, engage their audience, and improve their online presence, particularly in digital marketing and SEO (Search Engine Optimization). This strategy involves consistently updating and adding new content to a business's website, social media channels, and other marketing platforms.

- **Keeping Content Current:** ChatGPT can help in regularly updating website content, promotional materials, and other marketing texts to keep them fresh, relevant, and engaging.

Feedback Incorporation: This approach involves collecting, analyzing, and implementing feedback from various sources to improve marketing strategies and tactics. This is essential for small businesses to stay attuned to customer needs, preferences, and market trends.

- **Iterative Improvement:** Based on customer feedback and performance metrics, ChatGPT can assist in refining and iterating on marketing copy to continuously improve its effectiveness.

Using ChatGPT for crafting marketing copy offers a blend of creativity and efficiency. It allows for scalability in content creation and ensures that the marketing materials are not only persuasive and well-targeted but also consistently aligned with your company's brand voice and the audience's evolving needs. However, as always, it's essential to complement AI-generated content with personal oversight to ensure proper brand alignment, accuracy, and the nuanced understanding that only human expertise can provide.

SOCIAL MEDIA MANAGEMENT WITH CHATGPT

Social media platforms are vital for building brand awareness and engaging your audience in real-time. ChatGPT can assist in managing your social media presence on a multitude of platforms effectively.

- **Content Scheduling:** Use AI to schedule social media posts at optimal times, ensuring consistent and timely engagement with your followers.
- **Content Ideas:** Generate content ideas for your social media posts, including relevant hashtags and trending topics.
- **Community Engagement:** ChatGPT can assist in responding to comments and messages on your social media profiles and maintaining an active and responsive online presence.
- **Social Media Analytics:** Analyze social media data with AI to gain insights into the performance of your posts and campaigns, enabling data-driven decisions.
- **Content Reposting:** Repurpose and repost AI-generated content to keep your social media feeds fresh and engaging.

EMAIL MARKETING ASSISTANCE

Email marketing still remains a powerful tool for nurturing leads and retaining customers. ChatGPT can aid businesses in creating persuasive email subject lines and body content to improve email marketing campaigns' open and click-through rates.

- **Email Campaign Ideas:** Collaborate with AI to brainstorm ideas for email marketing campaigns, such as product launches, newsletters, and promotions.
- **Email Subject Lines:** Generate attention-grabbing subject lines with emojis that increase email open rates and encourage recipients to read your messages.
- **Newsletter Content:** Create informative and engaging newsletter content that provides value to your subscribers and builds brand loyalty.

- **Segmentation Strategies:** Use AI to segment your email list based on customer behavior, preferences, or demographics, ensuring personalized and relevant content.
- **Automated Follow-ups:** Implement automated follow-up sequences with ChatGPT to nurture leads and guide them through the sales funnel.

Leveraging ChatGPT in your content creation and marketing efforts can save time, enhance the quality of your content, and improve customer engagement. As we progress in this book, you'll discover more ways to harness AI for small business success, including data analysis, automation, and e-commerce strategies. Review the case study below on how a small business integrated ChatGPT for their marketing needs.

CASE STUDY: TRANSFORMING MARKETING STRATEGIES WITH CHATGPT

A Texas-based startup that focuses on innovative smart home devices struggled to gain traction in a saturated market, despite having groundbreaking products. Their marketing efforts were inconsistent and failed to effectively communicate the unique value proposition of their products.

Challenge: As a startup with limited resources (like most small businesses), this emerging tech giant desperately needed a cost-effective solution to enhance its marketing strategies. They required a tool that could consistently produce high-quality, engaging content to build brand awareness and drive customer engagement.

Solution: They utilized ChatGPT for content creation and marketing efforts. The prompts used and outcomes are below.

> Prompt: Write a blog post about the benefits of smart home devices in energy conservation.

> Outcome: ChatGPT generated a plethora of informative and engaging blog posts on topics like smart home technology, product usage tips, and industry trends.

Prompt: Create concise and compelling product descriptions for our new range of smart thermostats [uploaded data].

> Outcome: ChatGPT crafted persuasive and clear marketing copy for their website, product descriptions, and online advertisements.

Prompt: Develop a series of engaging tweets highlighting the unique features of our smart security cameras.

> Outcome: ChatGPT generated regular social media posts, including promotional content, interactive polls, and tech tips, to enhance online presence and engagement.

Prompt: Compose a personalized email for returning customers offering exclusive discounts on new products.

> Outcome: ChatGPT created personalized email marketing campaigns, targeting different segments of their customer base with tailored messages and offers.

Prompt: Generate a content calendar for the next quarter focusing on weekly blog posts, bi-weekly social media updates, and monthly email campaigns related to technology advancements and user tips.

> Outcome: A content calendar was created, outlining the schedule for blog posts, social media updates, and email campaigns.

Prompt: Track engagement metrics for our email campaigns and provide insights on the most effective content for our audience [included metrics].

> Outcome: Customer feedback was monitored to gauge the effectiveness of the content, and adjustments were made accordingly.

Prompt: Adjust the tone of our AI-generated content to more closely align with our brand ethos, particularly in ensuring that it is approachable yet authoritative in the field of technology.

Outcome: Efforts were made to ensure that all AI-generated content aligned with their brand voice and ethos.

Results: Within six months of integrating ChatGPT, this tech start-up experienced dynamic performance upgrades with its marketing efforts.

- **Increased Web Traffic:** There was an increase of 42.4% in website visitors, primarily attributed to engaging blog content and effective SEO.
- **Enhanced Social Media Presence:** Their social media following doubled, with a 62.5% increase in user engagement on platforms like Twitter, Instagram, and LinkedIn, all aligned with their target audiences.
- **Higher Email Open Rates:** Email marketing campaigns saw a 25% increase in open rates and a 16.7% increase in click-through rates.
- **Cost-Effective Marketing:** Overall marketing costs were reduced by 32% as the need for external content creation services decreased.

Conclusion: This situation showcases the profound impact of integrating AI in content creation and marketing strategies. By leveraging ChatGPT, this tech startup produced consistent, high-quality content that resonated with their audience. They also significantly enhanced its digital presence and increased customer engagement. This case study demonstrates the potential of AI tools like ChatGPT in empowering small businesses and startups to execute effective marketing strategies with limited resources.

ASK THE BOT: ADDITIONAL AI COMMANDS

Here are seven additional marketing prompts for small businesses using ChatGPT to generate engaging blog posts, craft persuasive marketing campaigns, manage social media, and assist with email marketing. Open up and log into ChatGPT on your mobile device or computer and enter the following queries.

I need fresh ideas on how ChatGPT can help me brainstorm engaging blog posts for [enter your business, website, or industry] and generate creative topics that resonate with my audience.

How can ChatGPT assist me in crafting persuasive marketing copy and developing a strategy that drives customer engagement and conversions on our current [or new] marketing campaign for [enter your business, website, or industry]?

Create engaging social media posts, captions, and content ideas that resonate with our followers and boost our online presence for [enter your business, website, or industry].

Provide insights on my email marketing campaigns' subject lines, content ideas, and strategies to increase open and click-through rates for [enter your business, website, or industry]'s current ad campaign.

Translate our ad campaign or social media post [copy and paste into ChatGPT] for [enter your business, website, or industry] into [specify a language].

Once I've created an email marketing campaign, how can ChatGPT help me analyze and optimize its performance for [enter your business, website, or industry]?

> What key metrics should I focus on, and how can I enhance the effectiveness of my email outreach for [enter your business, website, or industry]?

These prompts above will guide you and your small businesses in utilizing ChatGPT to enhance your content creation, marketing, and customer engagement efforts across various channels.

ADDITIONAL TRAINING

Since ChatGPT is continually evolving and each industry is unique, we have created extensive online training programs on our website, www.DynamicChatGPT.com, to further assist you in harnessing the power of AI for your small business.

CHAPTER 7
DATA ANALYSIS AND MARKET RESEARCH WITH CHATGPT

KNOWING your customers and target audience is fundamental in any business. Unfortunately, data analysis and market research are two things most small businesses spend little to no time or money on. Over the years, with all the companies we've owned—everything from telecom in the 1980s and 1990s to a significant health and life insurance brokerage in the first decade of the 2000s to an energy deregulation business in the 2010s to a publishing business currently, we found the process of data analysis and market research to be excruciating. Data analysis and market research are also very time-consuming and expensive.

However, with AI, these activities don't have to be for the small business owner. By automating the process using ChatGPT, we will demonstrate how you can have all the data necessary to make informed business decisions and leave nothing to chance. And it can all be done in the blink of an eye.

This chapter will explore how ChatGPT can be a valuable tool for data analysis and market research in your small business. From analyzing customer feedback to predicting market trends and optimizing pricing strategies, AI-powered data-driven insights can help you make informed decisions and stay competitive.

ANALYZING CUSTOMER FEEDBACK WITH CHATGPT

Understanding customer feedback is paramount for improving products and services. ChatGPT can assist you in analyzing customer feedback efficiently and help you make better informed business decisions.

- **Sentiment Analysis:** Utilize ChatGPT to analyze customer reviews, comments, and feedback sentiment. It can help you gauge customer satisfaction and identify areas that need improvement.
- **Feedback Categorization:** Automate the categorization of customer feedback into topics or themes. This allows you to identify recurring issues or trends in customer opinions.
- **Data Visualization:** Collaborate with ChatGPT to create visual representations of customer feedback data, such as word clouds or graphs, making interpreting and sharing insights easier.
- **Customer Insights:** ChatGPT can provide insights into customer preferences, pain points, and suggestions, helping you tailor your products or services to meet their needs.
- **Continuous Monitoring:** Set up automated processes with ChatGPT to continuously monitor and analyze incoming customer feedback, ensuring you stay responsive to changing customer sentiments.

MARKET TREND PREDICTION AND ANALYSIS

Staying ahead of market trends is necessary for small businesses. ChatGPT can assist in predicting and analyzing market trends.

- **Data Aggregation:** Collect data from various sources, including industry reports, social media, news articles, and more, using ChatGPT's data scraping capabilities.
- **Trend Identification:** Analyze the data to identify emerging trends and patterns in your industry or market segment.

- **Competitive Intelligence:** Use AI to track and analyze competitors' activities, product launches, and market strategies, helping you make informed decisions.
- **Predictive Analytics:** Leverage ChatGPT's predictive capabilities to forecast market trends and anticipate customer preferences or demand shifts.
- **Real-Time Updates:** Set up alerts and notifications with ChatGPT to receive real-time updates on market trends, ensuring you stay agile in response to changing conditions.

COMPETITOR ANALYSIS WITH AI INSIGHTS

Understanding your competitors is vital for maintaining a competitive edge. ChatGPT can provide valuable insights into your competitors.

- **Competitor Profiling:** Use AI to generate detailed profiles of your competitors, including their strengths, weaknesses, and market positioning.
- **Content Analysis:** Analyze competitors' content, such as blog posts, social media updates, and product descriptions, to identify their messaging strategies and content performance.
- **Pricing Strategies:** ChatGPT can assist in analyzing competitors' pricing strategies and suggest pricing adjustments to remain competitive.
- **Market Share Analysis:** Utilize AI to estimate competitors' market share and track changes over time, helping you identify growth opportunities.
- **SWOT Analysis:** Collaborate with ChatGPT to perform SWOT (Internal Strengths and Weaknesses and External Opportunities and Threats) analyses on your competitors, providing a comprehensive view of their market presence.

We perform an in-depth SWOT analysis for every new business we have entered. Each time, the analysis has produced new insights into additional markets or initiatives we need to take advantage of and address.

PRICING STRATEGIES AND OPTIMIZATION

Optimizing pricing strategies is one of the most essential foundational blocks for profitability. ChatGPT can assist in pricing analysis and optimization.

- **Dynamic Pricing:** Implement these strategies based on AI insights, allowing you to adjust prices in real-time based on market conditions and demand fluctuations.
- **Price Elasticity Analysis:** Utilize ChatGPT to analyze price elasticity in your market, helping you determine how changes in pricing will impact demand.
- **Competitive Pricing:** AI can assist in setting economical pricing by analyzing your competitors' pricing strategies and suggesting adjustments for you to gain a competitive edge.
- **Discount Strategies:** Collaborate with ChatGPT to develop effective discount strategies that maximize revenue without compromising profitability.
- **Price Testing:** Use AI to design and conduct price testing experiments, helping you identify optimal price points for your products or services.

By leveraging ChatGPT for data analysis and market research, you can gain valuable insights, stay informed about market trends, and make data-driven decisions that enhance your small business's competitiveness and profitability. As we continue this book, you'll discover additional ways AI can benefit your small business, from automation to e-commerce strategies and more.

CASE STUDY: STREAMLINING CUSTOMER FEEDBACK ANALYSIS WITH CHATGPT

A growing fast-casual urban restaurant chain is known for its innovative fusion cuisine and has experienced issues with efficient processing and acting upon customer feedback. With multiple locations and an active online presence, they received substantial feedback across various platforms, which was overwhelming their team.

Challenge: To effectively analyze and categorize the vast amount of customer feedback and identify key areas for improvement and customer preferences. They needed a system to quickly and efficiently process their data and provide actionable insights.

Solution: They decided to employ ChatGPT to analyze and categorize customer feedback. The prompts used and outcomes are below.

Prompt: Analyze the sentiment of recent customer reviews on social media and feedback forms to determine overall satisfaction levels and identify any recurring positive or negative trends [data entered].

Outcome: ChatGPT was programmed to analyze the tone of their customer reviews and comments from social media, emails, and feedback forms.

Prompt: Identify keywords and phrases frequently used in positive and negative feedback to better understand customer sentiment trends [data entered].

Outcome: They were able to better process customer feedback and added keywords to FAQs to improve overall customer satisfaction and identify negative or positive trends.

Prompt: Categorize customer feedback into themes such as food quality, service efficiency, ambiance, and pricing for targeted analysis.

Outcome: The AI was used to categorize feedback into themes such as food quality, service, ambiance, and pricing.

Prompt: Automatically tag and sort incoming customer feedback to streamline the review process [entered data].

Outcome: ChatGPT made the process easier for management to address specific issues and craft solutions by automatically tagging and sorting feedback for easy reference and analysis.

Prompt: Generate comparative visual data to show changes in customer feedback themes over time, helping identify improvement areas or successful initiatives.

Outcome: ChatGPT created word clouds and pie charts to depict common feedback themes and sentiments, making the data more accessible and understandable for both management and staff.

Prompt: Analyze customer responses to identify the most and least popular menu items, providing insights for potential menu adjustments and creating a training guide.

Outcome: The AI analyzed customer responses to provide insights into customer preferences and pain points, identified popular menu items or common service-related complaints, and created a rating and operational guide for this restaurant chain.

Prompt: Set up an alert system for immediate notification of significant shifts in customer feedback trends or the emergence of critical issues.

Outcome: ChatGPT was set up to continuously monitor incoming feedback, ensuring real-time analysis and responsiveness, analyzing incoming customer feedback, and providing regular updates and reports on trends and customer satisfaction levels to better inform management and staff.

Prompt: Integrate ChatGPT with customer evaluation channels, including social media, emails, and feedback forms, for centralized and efficient processing [included data].

Outcome They integrated ChatGPT with their various evaluation channels for centralized processing.

Prompt: Train ChatGPT with industry-specific language and details about their menu and service style to ensure accurate and relevant feedback analysis.

Outcome: The AI was trained with industry-specific language and their menu and service style for accurate analysis.

Prompt: Regularly update ChatGPT's knowledge base with new menu items, promotions, and service changes to keep the analysis current and relevant.

Outcome: Continuously updated ChatGPT's database with new menu items, promotions, and changes in service for up-to-date analysis.

Results: After implementing ChatGPT for six months, this fast-casual food chain experienced astonishing impacts on their customer service processes and satisfaction.

- **Enhanced Understanding of Customer Sentiment:** They experienced a 55% improvement in understanding customer sentiments and trends.

- **Streamlined Feedback Processing:** The time taken to process and categorize feedback was reduced by 77%.
- **Informed Decision-Making:** Insights from customer feedback led to menu adjustments and service improvements, resulting in a 26.3% increase in customer satisfaction scores.
- **Proactive Response to Feedback:** Real-time analysis allowed quicker responses to customer concerns, enhancing brand reputation.

Conclusion: The use of ChatGPT for analyzing customer feedback demonstrates the potential of AI in transforming the way businesses understand and respond to customer insights. By leveraging ChatGPT, this restaurant chain was able to efficiently process large volumes of feedback, gain actionable insights, and make informed decisions to enhance its customer experience. This case study highlights the importance of AI tools in managing and utilizing customer feedback effectively in the fast-paced restaurant industry.

ASK THE BOT: ADDITIONAL AI COMMANDS

Below are six prompts for small business owners on using ChatGPT to analyze customer feedback, predict market trends, conduct competitor analysis, and optimize pricing strategies. Open up and log into ChatGPT on your mobile device or computer and enter the following queries.

> How can I use ChatGPT to analyze this feedback effectively to identify trends, areas for improvement, and customer sentiment? Provide an analysis of [enter your business, website, or industry]'s customer feedback, reviews, and insights over [your platform of choice, i.e., Amazon, Google, etc.].

> How can ChatGPT help me analyze data and predict emerging trends in my industry or niche? Identify emerging trends and patterns in the [your] industry.

Create a SWOT analysis of [enter your business, website, industry, and all supporting data].

How can ChatGPT assist in conducting competitive analysis and providing recommendations for staying competitive? Compile a competitor analysis of [enter your business, website, or industry]. Supply insights into my competitors' strategies and performance.

How can ChatGPT assist in analyzing market dynamics, pricing trends, and optimizing our pricing structure to maximize profitability? List pricing strategies for my products or services.

Analyze [enter your business, website, or industry] competitors' pricing strategies and suggest pricing adjustments to gain a competitive edge in the [your] industry.

These prompts will help small business owners leverage ChatGPT to analyze customer feedback, stay informed about market trends, gain insights into competitors, and optimize their pricing strategies for success.

ADDITIONAL TRAINING

Since ChatGPT is continually evolving and each industry is unique, we have created extensive online training programs on our website, www.DynamicChatGPT.com, to further assist you in harnessing the power of AI for your small business.

CHAPTER 8
USING AI FOR AUTOMATING ADMINISTRATIVE TASKS

AS WE ALL KNOW, efficient, personable, and productive administrative assistants can do wonders for a small business. Now, imagine the heavens opening up and down floats the most ah-maz-ing administrative assistant on the planet. They have the largest task capacity you have seen—they can do the job of four or five people at a time without ever complaining. They are always joyful, have no personal issues, never call out sick, are always on time, and are productive twenty-four hours a day. Then you wake up from your dream.

But that is your reality with what AI through ChatGPT provides you—an assistant who knows everything you tell it, easily accomplishes tasks in seconds that would previously take you or your staff over a week to complete, is at your disposable via your mobile device and computer any time of day or night, and relieves you of the mountain of non-income-producing tasks that previously littered your desk.

In this chapter, we'll explore how AI can streamline administrative tasks for your business. From managing appointments and schedules to handling accounting, inventory management, and human resources, AI-powered automation can help you save time, reduce errors, and improve efficiency.

MANAGING APPOINTMENTS AND SCHEDULES

Efficient appointment and schedule management is important for businesses of all sizes. Everyone knows that missing or double booking an appointment questions the credibility and professionalism of not only the individual but also the organization. AI can assist in automating these administrative tasks.

- **Appointment Scheduling:** Implement AI-driven chatbots or virtual assistants to handle appointment scheduling. Customers can book appointments through a user-friendly interface, and the AI system can manage the calendar.
- **Automated Reminders:** Set up automated reminder systems that use AI to send appointment reminders to customers via email, SMS, or other channels.
- **Calendar Optimization:** AI can analyze your schedule to identify opportunities for optimizing appointments, reducing gaps, and maximizing productivity.
- **Appointment Analytics:** Use AI to analyze appointment data, identifying trends and patterns that can inform your business strategy.
- **Customized Availability:** AI can adapt to your business's unique scheduling needs, allowing you to set custom availability windows based on your preferences and capacity.

ACCOUNTING AND FINANCIAL REPORTING

Accurate financial management is vital for the success of your small business. Thankfully, AI can streamline accounting and financial reporting processes.

- **Expense Tracking:** Implement AI-based expense tracking systems that automatically categorize expenses and generate reports.

- **Invoice Generation:** Use AI to automate the creation and distribution of invoices, reducing manual data entry and errors.
- **Predictive Analytics:** AI can provide insights into financial trends, helping you make informed decisions about budgeting, investments, and resource allocation.
- **Fraud Detection:** Implement AI algorithms to identify irregularities and potential fraud in financial transactions.
- **Financial Reporting:** Automate generating financial reports, saving time and ensuring accuracy in your financial statements.

INVENTORY MANAGEMENT WITH AI

Efficient inventory management is critical for businesses that deal with physical products. AI can optimize inventory management processes.

- **Demand Forecasting:** AI can analyze historical data and market trends to predict future demand for your products, helping you maintain optimal inventory levels.
- **Inventory Reordering:** Automate reordering inventory when stock levels reach predefined thresholds, ensuring you never run out of your products.
- **Stock Optimization:** AI algorithms can optimize stock allocation, ensuring that high-demand items are readily available while minimizing overstocking of slow-moving products.
- **Real-Time Tracking:** Implement AI-driven tracking systems that provide real-time visibility into your inventory, allowing for better decision-making.
- **Supplier Insights:** AI can analyze supplier performance data to help you make informed sourcing and supplier relationship decisions.

STREAMLINING HR AND EMPLOYEE ONBOARDING

Human resources and employee onboarding are critical functions for any business. AI can simplify these processes.

- **Resume Screening:** Use AI to automate the screening of job applications and resumes, identifying candidates who meet your criteria.
- **Interview Scheduling:** AI-powered chatbots can schedule interviews with candidates, eliminating the back-and-forth communication between HR and applicants.
- **Employee Onboarding:** Automate the onboarding process with AI, providing new hires access to much-needed resources, training materials, and information.
- **HR Analytics:** ChatGPT can analyze HR data to identify employee performance, engagement, and retention trends, helping you make informed HR decisions as well as creating engaging interview questions.
- **Compliance Management:** Implement AI-driven systems to track and ensure compliance with labor laws and regulations, reducing the risk of legal issues.

Incorporating AI into your administrative tasks can reduce the time and effort spent on manual processes, allowing you to focus on strategic initiatives that drive your business forward. As we continue through this book, you'll discover how AI can benefit your small business, including e-commerce strategies, customer service, and more.

CASE STUDY: REVOLUTIONIZING ADMINISTRATIVE TASKS WITH AI

A small accounting firm in Oregon, which offers tax preparation, bookkeeping, and financial consulting services, was struggling. Despite its reputation for excellent customer service, it faced problems managing its growing administrative workload. Scheduling appointments, managing inventory, and handling accounting and HR processes were becoming increasingly time-consuming, impacting staff productivity.

Challenge: This accounting firm needs an efficient way to handle administrative tasks without compromising accuracy or customer service quality. Like most small businesses with a limited team and a tight budget, they sought a solution to automate routine tasks, reduce errors, and improve overall efficiency.

Solution: Implement AI-powered tools to streamline their administrative processes. The prompts used and outcomes are below.

Prompt: Automate appointment scheduling, provide confirmation, and create reminder notifications to clients.

Outcome: Reduced scheduling conflicts and no-shows and improved client satisfaction using an AI-powered scheduling system integrated with the firm's website.

Prompt: Automatically categorize expenses and generate financial reports monthly.

Outcome: More accurate financial reporting, reduced time spent on manual bookkeeping utilizing an AI software for real-time bookkeeping and financial reporting.

Prompt: Monitor office supply levels and automatically reorder items when low.

Outcome: They optimized inventory levels and reduced instances of supply shortages with their new inventory management system using AI to track and reorder supplies.

Prompt: Automate payroll processing and maintain up-to-date employee records.

Outcome: Streamlined payroll process and decreased administrative errors with an HR platform for payroll processing and employee data management.

Results: After six months of ensuring AI tools were compatible with existing software, training their staff with workshops to familiarize staff with the new AI systems, and regular monitoring and adjusting AI performance, they experienced a more administratively efficient office.

- **Increased Efficiency**: Reduced time spent on administrative tasks by 46.5%.
- **Error Reduction:** Decreased errors in bookkeeping and payroll processing by 51%.
- **Improved Inventory Management:** Achieved a 31.6% cost saving on office supplies through efficient inventory management.
- **Enhanced Customer Service:** Client appointment scheduling satisfaction increased by 66.8%.

Conclusion: The accounting firm's success demonstrates how small businesses can significantly benefit from implementing AI to manage administrative tasks. Using AI in appointment scheduling, bookkeeping, inventory, and HR processes not only saved time and reduced errors but also allowed their team to focus more on client service and core business activities. This case study highlights the potential of AI-powered automation in enhancing the efficiency and effectiveness of administrative operations in small businesses.

ASK THE BOT: ADDITIONAL AI COMMANDS

To assist you in applying the administrative knowledge you learned from this chapter, open up and log into ChatGPT on your mobile device or computer and enter the following queries.

How do I use AI chatbots to handle appointment scheduling through a user-friendly interface and manage my calendar?

How do I use automated AI reminder systems to send appointment reminders to customers via email, SMS, or other channels?

> How do I implement AI-based expense tracking systems that automatically categorize expenses and generate reports for [enter your business, website, or industry]?

> How can AI automate [enter your business, website, or industry]'s onboarding process, providing new hires access to essential resources, training materials, and information?

> Provide twenty interview questions for a new employee for [enter your business, website, or industry].

> How do I implement AI-driven systems to track and ensure compliance with labor laws and regulations, reducing the risk of legal issues for [enter your business, website, or industry]?

These prompts will help you and your small businesses leverage ChatGPT to offload administrative and human resources tasks such as appointment scheduling, client reminders, expense tracking, and employee onboarding.

ADDITIONAL TRAINING

Since ChatGPT is continually evolving and each industry is unique, we have created extensive online training programs on our website, www.DynamicChatGPT.com, to further assist you in harnessing the power of AI for your small business.

CHAPTER 9
USING CHATGPT IN E-COMMERCE

E-COMMERCE NEEDS NO INTRODUCTION. It covers all commercial transactions conducted electronically on the internet. We all know we can't do business without it—from receiving payments via money transfer apps to generating invoices, e-commerce is foundational, especially for small businesses. And if we aren't selling online in some capacity, we aren't maximizing our organization's potential.

In this chapter, we'll explore how ChatGPT can revolutionize the e-commerce landscape. From personalized product recommendations to crafting product descriptions and managing inventory, ChatGPT offers a range of solutions to enhance your e-commerce business.

PERSONALIZED PRODUCT RECOMMENDATIONS

Personalization is a critical driver of e-commerce success, and ChatGPT can play a pivotal role in delivering tailored product recommendations to your customers.

- **User Profiling:** Utilizing ChatGPT to scrutinize patterns in user interactions can be a game-changer for businesses seeking to understand their clientele on a deeper level. By examining the nuances in customer queries, purchases, and feedback, ChatGPT can help in crafting highly personalized user profiles that inform more targeted marketing strategies and product development.

- **Recommendation Engines:** Integrate ChatGPT with recommendation engines to provide personalized product suggestions to customers, increasing the likelihood of conversion.
- **Dynamic Content:** Implement AI-driven personalization copy on your website, displaying product recommendations that change based on the user's browsing history and interests.
- **Chatbot Recommendations:** Use AI chatbots powered by ChatGPT to engage with customers in real-time, offering personalized product recommendations during conversations.
- **Email Marketing:** Use ChatGPT to generate personalized email content and electronic marketing campaigns, including product recommendations.

CREATING PRODUCT DESCRIPTIONS AND REVIEWS

Compelling product descriptions and reviews are vital for boosting sales. ChatGPT can assist in creating high-quality content for your e-commerce site.

- **Product Descriptions:** Automate the generation of product descriptions using ChatGPT, ensuring consistency and clarity in your product listings.
- **Review Generation:** Encourage customers to leave reviews and use AI to analyze the data and generate summaries highlighting key insights and sentiments.
- **SEO Optimization:** Collaborate with ChatGPT to optimize product descriptions and reviews for search engines, improving the discoverability of your products.

- **Multilingual Content:** Use AI to translate product descriptions and reviews into multiple languages, expanding your global reach.
- **Bulk Content Generation:** Quickly generate product descriptions and reviews for extensive inventories, saving time and effort.

INVENTORY MANAGEMENT AND SALES FORECASTING

Efficient inventory management and accurate sales forecasting are the basis for e-commerce success in businesses. ChatGPT can streamline these processes.

- **Inventory Optimization:** Leverage ChatGPT to analyze sales data and predict inventory needs, ensuring you have the right products in stock without overstocking.
- **Demand Forecasting:** Use AI to forecast future sales trends and demand fluctuations, helping you make informed procurement decisions.
- **Real-Time Inventory Tracking:** Implement AI-driven systems that provide real-time visibility into your inventory levels, enabling better decision-making.
- **Sales Reporting:** Generate automated sales reports using ChatGPT, providing insights into your e-commerce performance.
- **Supplier Communication:** Automate communications with suppliers based on inventory data, streamlining the reordering process.

CUSTOMER RETENTION AND UPSELLING STRATEGIES

Building customer loyalty and increasing customer lifetime value are two building blocks for e-commerce success. ChatGPT can help with customer retention and upselling.

- **Personalized Customer Engagement:** Use AI chatbots to engage with customers personally, offering tailored recommendations and promotions based on their purchase history.
- **Loyalty Programs:** Implement AI-powered loyalty programs that reward customers for repeat purchases, with ChatGPT assisting in program management.
- **Customer Feedback Analysis:** Analyze customer feedback with ChatGPT to identify areas for improvement and enhance customer satisfaction.
- **Upselling and Cross-selling:** AI-driven upselling and cross-selling recommendations can be integrated into the shopping experience, increasing average order value.
- **Retention Email Campaigns:** Generate personalized email content for retention campaigns, such as product replenishment reminders and exclusive offers.

Leveraging ChatGPT in your e-commerce operations can significantly enhance customer experiences, streamline processes, and drive revenue growth. As we continue through this book, you'll discover more ways AI can benefit your small business, from customer service to data analysis and beyond.

CASE STUDY: ENHANCING E-COMMERCE EFFICIENCY WITH CHATGPT

An emerging online fashion retailer based in Florida specializing in contemporary apparel is struggling with the demands of online retail despite having a trendy and appealing product line. They are also experiencing difficulties scaling their e-commerce efficiency to provide personalized shopping experiences, manage inventory effectively, and create engaging product descriptions.

Challenge: The issues faced by this fashion retailer are twofold: firstly, to improve the online shopping experience for their customers through personalization, and secondly, to streamline their inventory management and content creation processes. They needed a solution that could handle these aspects efficiently and cost-effectively.

Solution: Integrate ChatGPT into their e-commerce platform, focusing on several key functionalities. The prompts used and outcomes are below.

Prompt: Analyze customer browsing and purchase histories to offer personalized product recommendations.

> Outcome: ChatGPT was trained to analyze customer browsing and purchase histories to offer personalized product recommendations.

Prompt: Create an interactive chatbot interface on our website to provide personalized fashion advice based on customer preferences.

> Outcome: Customers received tailored fashion advice and suggestions through a chatbot interface on their website.

Prompt: Generate unique and compelling product descriptions based on inputted product features and specifications.

> Outcome: ChatGPT generated unique and compelling product descriptions for the online catalog by inputting product features and specifications into ChatGPT to create descriptive and persuasive text.

Prompt: Track inventory levels and predict restocking needs based on sales data analysis [data entered].

> Outcome: The AI tool was also set up to track inventory levels, predict restocking needs, and identify slow-moving items by analyzing sales data to provide restocking alerts and inventory reports.

Prompt: Automate responses to routine customer queries regarding order status, product information, and return policies.

Outcome: ChatGPT handled routine customer queries regarding order status, product information, and return policies using an automated chat service on the website and social media platforms.

Results: ChatGPT was integrated with its online platform, including its product database and customer service channels. AI was trained with specific language and knowledge pertinent to the fashion industry and product range. After four months, this online fashion retailer experienced an increase in venue, customer engagement, and

- **Increased Sales through Personalization:** A 29.1% increase in sales was attributed to personalized product recommendations.
- **Improved Customer Engagement:** Product page engagement rates improved by 39.9% due to compelling product descriptions.
- **Efficient Inventory & Forecasting Management:** Stock levels were optimized, reducing overstock by 28% and understock inventory by 16.3%.
- **Enhanced Customer Service:** Customer service efficiency increased, with a 54.4% reduction in response time for customer inquiries.

Conclusion: The integration of ChatGPT revolutionized this online fashion retailer's approach to e-commerce. By utilizing AI for personalized recommendations, content creation, and inventory management, they were able to provide a superior shopping experience, optimize their operations, and increase sales. This case study demonstrates the significant impact that AI, particularly ChatGPT, can have in enhancing the various aspects of an e-commerce business, from customer interaction to backend management.

ASK THE BOT: ADDITIONAL AI COMMANDS

To assist you in using ChatGPT to apply the e-commerce knowledge from this chapter on personalized product recommendations, creating product descriptions and reviews, inventory management, sales forecasting, and customer retention and upselling strategies, enter the command prompts below into ChatGPT on your mobile device or computer.

How can ChatGPT help us implement a recommendation engine that tailors suggestions based on individual preferences for [enter your business, website, or industry]? Provide improvements to our e-commerce website by offering personalized product recommendations to customers.

How can ChatGPT assist in generating persuasive and detailed product descriptions highlighting our offerings' unique features and benefits for [enter your business, website, or industry]? Refresh our product descriptions to attract more customers.

How can ChatGPT help us optimize [enter your business, website, or industry]'s inventory levels, predict demand, and ensure we have the right products in stock?

How can ChatGPT assist in analyzing historical sales data and market trends to provide insights and predictions that guide our sales strategies for [enter your business, website, or industry]? Improve our sales forecasting accuracy.

How can ChatGPT help us develop effective strategies, such as personalized offers, loyalty programs, and follow-up communications for [enter your business website, or industry]? Increase our customer retention and upselling capacity to existing customers.

These prompts will help you and your small businesses leverage ChatGPT to enhance their product offerings, improve inventory management, and implement effective customer retention and upselling strategies.

ADDITIONAL TRAINING

Since ChatGPT is continually evolving and each industry is unique, we have created extensive online training programs on our website, www.DynamicChatGPT.com, to further assist you in harnessing the power of AI for your small business.

CHAPTER 10
INTEGRATING CHATGPT WITH YOUR WEBSITE AND APPS

THIS CHAPTER on AI integration into your company's website can seem a bit scary. If you're like thousands of other business owners, you might have had someone else create your website. You may have no idea how to access your website to add features. Don't worry.

In this chapter, we'll delve into seamlessly integrating ChatGPT into your website and mobile apps. From embedding ChatGPT to utilizing APIs for custom solutions, we'll explore creating a cohesive and user-friendly experience while addressing security and privacy considerations. And we will do this at a rudimentary level so it's easy to understand.

EMBEDDING CHATGPT ON YOUR WEBSITE

Integrating ChatGPT into your website can enhance user engagement and provide real-time support. The compatibility of ChatGPT with a specific chat widget or platform may depend on the integration options provided by that widget or platform, as well as the technical capabilities of your development team. It's essential to check the documentation and capabilities of the chat widget or platform you plan to use and consider your project's specific goals and requirements. Here's how to get started.

- **Chat Widget Integration:** Choose a chat widget compatible with ChatGPT, such as an AI-powered chatbot. Many platforms offer easy-to-install widgets that allow you to add ChatGPT to your website without extensive coding and allow integration with third-party chat tools.
- **User-Friendly Placement:** Determine the most user-friendly location for the chat widget, such as the bottom corner of your website or a dedicated "Contact Us" page.
- **Customizable Appearance:** Customize the appearance of the chat widget to match your website's branding, message, and design. This ensures a cohesive user experience.
- **Conversation Flow:** Define and chart the conversation flow for ChatGPT on your website. Consider the types of inquiries it will handle and the responses it should provide.
- **User Guidance:** Include clear instructions or prompts to guide users on how to interact with ChatGPT effectively. For example, you can instruct users to type "help" for assistance.

If you are unfamiliar with chatbots and installing them on your website, contact your web developer or check out our website, www.DynamicChatGPT.com, for additional information.

CHATGPT APIS FOR CUSTOM SOLUTIONS

Using ChatGPT APIs (application programming interfaces) allow for greater flexibility and customization in your integration efforts. API is a software interface that allows two or more computers to communicate. Below are important aspects to consider as you leverage ChatGPT APIs for your custom solutions.

- **API Access:** Obtain access to ChatGPT's APIs through the OpenAI platform. This may involve registering your application and obtaining API keys.
- **Development Resources:** Utilize your development team or hire developers experienced in working with APIs to implement ChatGPT into your website or app.

- **Custom Chat Experiences:** Build custom chat experiences that align with your business objectives. You can develop specialized chatbots or virtual assistants tailored to your industry or niche.
- **Scalability:** Consider the scalability of your integration. Ensure that the API-based solution can handle increasing user interactions and scale as your business grows.
- **Testing and Optimization:** Continuously test and optimize your custom ChatGPT integration to enhance its performance and responsiveness to user inquiries.

MOBILE APP INTEGRATION

Integrating ChatGPT into your mobile apps can enhance user engagement and provide on-the-go support. Here's how to integrate ChatGPT into your mobile apps effectively.

- **Mobile-Optimized Chat:** Ensure the chat interface is optimized for mobile devices, providing a seamless user experience.
- **In-App Integration:** Embed ChatGPT directly within your mobile app, allowing users to access it without navigating to a separate web page.
- **Offline Functionality:** Consider implementing offline functionality that allows users to access predefined responses or frequently asked questions without an internet connection.
- **Push Notifications:** Implement push notifications to proactively engage users with ChatGPT, such as sending product recommendations or updates.
- **Cross-Platform Compatibility:** Ensure compatibility across various mobile platforms (iOS, Android) and screen sizes to reach a broader audience.

SECURITY AND PRIVACY CONSIDERATIONS

Integrating ChatGPT into your website and apps comes with security and privacy considerations that need to be addressed before you and your team move ahead with implementation.

- **Data Encryption:** Ensure that all data transmitted between users and ChatGPT is encrypted to protect sensitive information.
- **User Data Handling:** Clearly communicate how user data is handled, stored, and used within your chat integration, adhering to data protection regulations.
- **User Consent:** This needs to be obtained for all data collection and usage, especially if you plan to gather user information for personalization.
- **Authentication:** Implement user authentication mechanisms to prevent unauthorized access to sensitive features or data.
- **Compliance:** Stay informed about privacy regulations and ensure that your ChatGPT integration complies with data protection laws relevant to your region or industry.

Integrating ChatGPT into your website and apps can provide valuable assistance to your customers and enhance their overall experience. By following best practices for integration and addressing security and privacy concerns, you can create a safe, efficient, and user-friendly environment that sets your business apart. As we continue through this book, you'll discover more ways that AI can benefit your small business, from automation to marketing strategies and beyond.

CASE STUDY: SIMPLIFYING CHATGPT INTEGRATION FOR ENHANCED CUSTOMER EXPERIENCE

A small but thriving online grocery store in Colorado aimed to provide a convenient and personalized shopping experience for its customers. However, handling customer queries efficiently and providing instant support on their website and mobile app proved problematic.

Challenge: This online grocer needed a solution that could offer real-time assistance to their customers for queries like product availability, recipe suggestions, and order tracking. They wanted to integrate a user-friendly AI solution that did not compromise customer security and privacy.

Solution: The store integrated ChatGPT into its website and mobile app. The process was kept simple and straightforward to ensure ease of use for both customers and their teams. The prompts used and outcomes are below.

Prompt: Integrate a non-intrusive yet easily accessible ChatGPT chat widget in the bottom right corner of the website.

Outcome: A chat widget powered by ChatGPT was added to the bottom right corner of the website. This widget was designed to be non-intrusive but easily accessible using a simple ChatGPT plugin compatible with their website platform, ensuring quick and hassle-free integration.

Prompt: Embed a ChatGPT API into the mobile app to provide real-time assistance, featuring it prominently in its main menu.

Outcome: They experienced consistent user experience across platforms, with increased engagement and positive feedback on the mobile app's chat. By integrating a ChatGPT API into their existing mobile app, they provided functionality similar to that on the website. They also worked with their app developers to include a chat feature in their main menu for easy accessibility.

Prompt: Customize ChatGPT to reflect our brand voice of exceptional customer care and personalized shopping experience to respond more accurately to specific customer service scenarios. [They entered FAQs and information about their grocery products and services].

Outcome: ChatGPT provided tailored responses to align with the grocer's voice and specific customer service scenarios, such as order inquiries and product questions, enhancing customer satisfaction and brand loyalty.

Prompt: Configure ChatGPT to comply with data protection regulations and avoid storing sensitive customer information.

Outcome: Maintained high customer data security and privacy standards, with no breaches reported.

Prompt: Configure ChatGPT to avoid storing sensitive customer information and direct more sensitive queries, like those on customer accounts, to human representatives.

Outcome: Maintained high customer data security and privacy standards, with no violations reported, as ChatGPT complied with data protection regulations and did not store sensitive customer information.

Prompt: Design the chat interface to be user-friendly for all demographics, with clear options and simple language, and include a feature where customers could rate their interaction with the AI for continuous improvement.

Outcome: Positive customer feedback on the ease of use and clarity of the chat interface. They designed the chat interface to be intuitive and easy to use for all customer demographics, with clear options and simple language.

Prompt: Conduct beta testing with a group of customers and gather feedback for improvements.

Outcome: Identified and rectified minor issues during the beta testing phase, leading to a smoother full rollout.

Prompt: Train our staff on the functionality of ChatGPT and its monitoring and updating processes.

Outcome: Provided basic training to the team on how the ChatGPT integration worked and how to monitor and update its functionality. They also empowered staff to effectively manage and update the AI system, ensuring its ongoing efficacy.

Results: After three months of implementation, this online grocery store experienced significant and enhanced customer experiences while maintaining the security and privacy of its users' data.

- **Improved Customer Engagement:** A 59% increase in customer interactions was noted, with positive feedback on the ease of use.
- **Reduced Response Time:** The average response time for online customer queries decreased by 72.7%.
- **Enhanced User Experience:** Customers appreciated the 24/7 availability of assistance on both the website and mobile app.
- **Maintained Security and Privacy:** No security or privacy breaches were reported, reflecting the effectiveness of the measures implemented.

Conclusion: The integration of ChatGPT into this online grocer's digital platforms demonstrates how AI can be seamlessly incorporated to enhance customer service. By focusing on a user-friendly design and addressing security and privacy concerns, they were able to provide an efficient and safe shopping experience. This case study exemplifies the potential of AI tools like ChatGPT in augmenting customer interaction and service in the e-commerce sector, even for businesses with limited technical resources.

ASK THE BOT: ADDITIONAL AI COMMANDS

To assist you in applying the knowledge on how to integrate ChatGPT with a variety of chat widgets and platforms to enable real-time interactions while ensuring security and privacy, open up and log into ChatGPT on your mobile device or computer and enter the following queries.

> Create a list of website chat widgets, messaging apps, custom chatbots, API Integration, and mobile apps compatible with ChatGPT for [enter your business, website, or industry].

> How can I seamlessly integrate ChatGPT into our small business website to provide real-time customer support and engage with website visitors effectively for [enter your business, website, or industry]?

> We want to enhance the user experience of our mobile app by integrating ChatGPT. What steps are involved in integrating ChatGPT into our mobile app, and what platforms does ChatGPT work best with [enter your business and industry]?

> Security and privacy are paramount for our business and customers. What security measures and best practices should we follow when integrating ChatGPT to protect sensitive data and user privacy for [enter your business, website, or industry]?

> How does ChatGPT handle data encryption and compliance with industry regulations? What steps should we take to ensure data security and compliance when using ChatGPT in our operations for [enter your business, website, or industry]?

These queries will help you as a small business owner gather important information and considerations when integrating ChatGPT into your online presence and applications while prioritizing security and customization.

ADDITIONAL TRAINING

Since ChatGPT is continually evolving and each industry is unique, we have created extensive online training programs on our website, www. DynamicChatGPT.com, to further assist you in harnessing the power of AI for your small business.

CHAPTER 11
FUTURE TRENDS IN AI
FOR SMALL BUSINESSES

WHEN KIM WORKED FOR GARTNER, an IT research and consulting firm, in the late 1990s, she had knowledge and access to the hottest newest tech before it reached the general market. Wouldn't it be wonderful if we all had that opportunity or omniscience? We could have the foresight to buy stocks before they skyrocketed. Or perhaps predict future business trends so we can get ahead of the curve and not be left in the dust by our competitors? That's precisely what we, as business owners, now have access to. We are on the cusp of the next technology revolution with AI and we need to make sure we are leveraging ChatGPT to its fullest potential for our business.

By now, you realize the impact AI and ChatGPT can have on your revenues. It's like having access to the top business consultants in your industry and having them tailor a specific solution to meet your immediate needs. What we've shared with you through this book enables you to put the power of this revolutionary technology in your hands. Now, it's time to look forward and get ahead of the growth curve for your industry. AI is rapidly changing and we need to keep up with it! A solid foundation will allow you to take advantage of this as AI expands over the next decade.

In this chapter, we'll explore AI's exciting and evolving landscape and its potential impact on small businesses. As Artificial Intelligence continues to evolve rapidly, small business owners must stay informed about the future trends and opportunities it provides.

THE EVOLVING ROLE OF AI IN BUSINESS

- **AI as a Business Driver:** AI is transitioning from a support tool to a critical business driver. Small businesses will increasingly rely on AI to gain a competitive edge, improve efficiency, and make data-driven decisions.
- **Hyper-Personalization:** AI will enable hyper-personalized customer experiences. Businesses will use AI to tailor products, services, and marketing campaigns to individual customer preferences, increasing customer loyalty and engagement.
- **AI in Decision-Making:** AI-driven analytics will play a more significant role in strategic decision-making. Businesses will use Artificial Intelligence to analyze complex data sets, identify trends, and make predictions that guide critical choices.
- **Automation of Knowledge Work:** Beyond repetitive tasks, AI will increasingly automate knowledge work, such as contract analysis, content generation, and even legal research, enabling small businesses to operate more efficiently.
- **AI in Supply Chain and Inventory Management:** AI will be pivotal in optimizing supply chain operations. Predictive analytics will help small businesses forecast demand accurately, reduce inventory costs, and improve delivery times.

UPCOMING AI TECHNOLOGIES AND OPPORTUNITIES

- **Natural Language Processing (NLP):** NLP advancements will enable more sophisticated chatbots and virtual assistants capable of understanding and responding to complex customer queries. Small businesses can use NLP to enhance customer support and communication.

- **AI-Powered Predictive Analytics:** Predictive analytics will become more accessible to small businesses. AI-driven forecasting models will help optimize inventory management, marketing campaigns, and financial planning.
- **AI in Marketing Automation:** AI will revolutionize marketing by automating content generation, personalizing marketing messages, and optimizing ad targeting. Small businesses can improve their marketing ROI through AI-powered campaigns.
- **AI in Healthcare:** AI-driven telemedicine solutions will continue to grow, providing small healthcare businesses with opportunities to expand their reach and offer remote services. AI can also enhance medical diagnostics and patient care.
- **AI-Enhanced Cybersecurity:** As cyber threats evolve, AI will play a crucial role in cybersecurity. Small businesses can use AI to detect and respond to security threats more effectively, protecting sensitive data and systems.
- **AI-Driven Sustainability Solutions:** AI will help small businesses adopt more sustainable practices by optimizing energy usage, reducing waste, and identifying eco-friendly operational alternatives.
- **AI for Enhanced Customer Insights:** AI-driven analytics tools will provide deeper insights into customer behavior and preferences, enabling small businesses to create more targeted marketing strategies and product offerings.
- **AI in Legal and Compliance:** Small businesses can leverage AI for legal research, contract analysis, and compliance management, reducing legal costs and ensuring regulatory adherence.

As AI evolves, small businesses embracing these trends and technologies will gain a competitive advantage. However, it's essential to approach AI adoption thoughtfully, considering your business's specific needs and challenges. Small businesses can thrive in an AI-driven future by staying informed and strategically implementing AI.

CASE STUDY: NAVIGATING THE EVOLVING AI LANDSCAPE WHILE EMBRACING MODERN TECHNOLOGY

A vintage record shop in Nashville that specializes in vinyl records and nostalgic music memorabilia is known for its classic charm. This business struggled to stay relevant in an increasingly digital music market while preserving its unique appeal.

Challenge: The primary conflict for this classic record store was to integrate modern technology without losing its vintage essence. They needed to stay informed of evolving AI trends and opportunities to enhance their business operations and customer experience.

Solution: Embarked on a journey to embrace AI technologies while maintaining its core identity, this retailer used the following ChatGPT prompts and experienced the subsequent outcomes.

Prompt: Develop a system to recommend music based on individual customer preferences, past purchases, and browsing habits. Also, provide steps to integrate the recommendation system into the online store and in-shop digital kiosks, ensuring user-friendliness and accessibility [entered data].

Outcome: They incorporated this system into their online store and in-shop digital kiosks. Customers received highly personalized music suggestions, resulting in increased loyalty and sales, with a notable improvement in customer satisfaction.

Prompt: Use AI to manage the vinyl records inventory, including tracking stock levels and predicting future demand trends based on sales data. Create AI algorithms to analyze sales data to forecast which genres or artists were gaining popularity.

Outcome: Utilized AI for managing their extensive inventory of vinyl records, tracking stock levels, and predicting future demand trends. They were able to optimize stock levels, reduce overstock and understock situations by 30%, and better align their inventory with customer demand.

Prompt: Leverage AI to analyze customer data and create targeted marketing campaigns for different segments of music enthusiasts.

Outcome: AI analyzed customer data to create specific marketing campaigns for different customer segments. Digital marketing campaigns became more effective, with a 20% increase in online sales attributed to more personalized promotions.

Prompt: Provide emerging AI technologies relevant to the retail and music industry. Include workshops, webinars, and tech networks. Also, create a list of AI newsletters, online forums, and online groups focused on retail marketing that is best for our business [entered business name and industry].

Outcome: Management regularly attended workshops and webinars on emerging AI technologies and participated in local small business tech networks. They remained at the forefront of AI adoption in their niche, continuously improving their use of AI in business operations.

Prompt: Ensure that the integration of AI technology complements the store's retro theme, with digital kiosks designed to match the vintage aesthetic.

Outcome: Balanced AI with efforts to maintain the store's classic ambiance, ensuring that technology complemented rather than overtook the retro experience. The successful integration of modern AI technology, without detracting from the store's nostalgic charm, was appealing to a wide range of customers. Digital kiosks were designed with a retro aesthetic. blending seamlessly with the store's vintage theme.

Prompt: Provide customer engagement ideas to gather feedback on new AI implementations to ensure they enhance the shopping experience without detracting from the store's charm.

Outcome: Valuable customer insights were obtained, leading to further refinements in the AI system for even better alignment with customer preferences.

Prompt: Create regular training sessions for staff on the use of AI tools in daily operations and customer interactions.

Outcome: Staff and management were more confident and effective in using AI tools, contributing to an overall smoother operation and enhanced customer service. They conducted ongoing training to keep everyone updated on AI tools and how to use them effectively in daily operations.

Results: Six months post-implementation, this vintage record store revamped its inventory processes and increased customer satisfaction while balancing the amalgamation of AI and its classic roots.

- **Enhanced Client Experience:** Customers enjoyed the personalized music recommendations, leading to increased customer loyalty and sales.

- **Efficient Inventory Management:** Stock levels were optimized, reducing overstock and understock situations by 36%.
- **Improved Marketing ROI:** Digital marketing campaigns became more effective, resulting in a 21.9% increase in online sales.
- **Balanced Modern and Vintage Appeal:** The shop successfully integrated AI without compromising its nostalgic appeal, attracting both new and loyal customers.

Conclusion: This vintage record store's journey in integrating AI showcases how small businesses can navigate the evolving landscape of Artificial Intelligence. By staying informed about AI trends and strategically implementing technology, they managed to enhance their business operations while preserving their unique identity. This case study highlights the importance of balancing technology adoption with core business values, a crucial consideration for small businesses in a rapidly changing digital world.

ASK THE BOT: ADDITIONAL AI COMMANDS

Below are five inquiries for small businesses seeking to use ChatGPT to predict future trends in AI and explore its potential applications, including AI as a business driver, hyper-personalization, AI in decision-making, automation of knowledge work, and supply chain management. Open up and log into ChatGPT on your mobile device or computer and enter the following queries.

> How can [enter your business, website, or industry] leverage ChatGPT and AI technologies to predict how AI will drive our business in the future? What emerging trends should we be aware of to stay competitive?

We're interested in providing hyper-personalized experiences to our customers for [enter your business, website, or industry]. How can ChatGPT help us predict and implement AI-driven hyper-personalization strategies effectively?

As we integrate AI into our decision-making processes, what insights can we gain from using ChatGPT to predict future trends and challenges in AI-driven decision-making for [enter your business, website, or industry]?

We want to automate knowledge work and streamline our operations. What trends should we consider when using ChatGPT to predict the future of AI in knowledge work automation for [enter your business, website, or industry]?

How can ChatGPT assist us in predicting future trends and innovations in AI for supply chain management for [enter your business, website, or industry]? What insights can we gather to enhance our supply chain operations?

These inquiries will help small businesses explore the potential of ChatGPT and AI in predicting and adapting to future trends in various aspects of their operations and industry.

ADDITIONAL TRAINING

Since ChatGPT is continually evolving and each industry is unique, we have created extensive online training programs on our website, www.DynamicChatGPT.com, to further assist you in harnessing the power of AI for your small business.

CHAPTER 12
ETHICAL CONSIDERATIONS AND RESPONSIBLE AI USAGE

IN 2002'S *SPIDER-MAN* MOVIE, Uncle Ben gave Peter Parker an ominous warning, "With great power comes great responsibility," referring to the incredible strength at Peter's disposal and the weight of the obligation accompanying such power. While AI and ChatGPT will not lead to robots taking over Earth, there are many ethical and moral obligations to consider when using such a powerful technology.

In this chapter, we'll explore the critical aspects of ethical considerations and responsible Artificial Intelligence usage. As small businesses increasingly integrate AI into their operations, it becomes paramount to ensure that AI is employed responsibly, respects privacy, and fosters trust among customers and stakeholders.

ENSURING FAIR AND ETHICAL AI PRACTICES

- **Bias Mitigation:** Small businesses must proactively identify and mitigate biases in AI algorithms. This includes reviewing data sources, training models on diverse data sets, and regularly evaluating AI systems for fairness.
- **Transparency:** It's vital to make AI processes transparent to stakeholders and customers and to explain how AI benefits them and addresses their needs. As business owners, we

need to communicate how AI is used, the data it relies on, and the decision-making processes it follows.

- **Ethical Guidelines:** Establish ethical guidelines for AI usage within your organization. Hopefully, you have an employee or business process manual—if you don't, have AI create one for you! These guidelines can help steer decisions and ensure AI aligns with your company's values.
- **Regular Audits:** Conduct periodic audits of AI systems to assess their ethical impact. This includes examining outcomes, unintended consequences, and feedback from users.
- **Diversity and Inclusion:** Encourage diversity in AI development teams. Diverse teams can help identify and address biases that may be overlooked in homogeneous groups.

DATA PRIVACY AND SECURITY

- **User Consent:** Ensure that users provide informed consent when collecting data for AI usage. Clearly explain how their data will be used and allow them to opt in or out as necessary.
- **Data Encryption:** Protect user data with robust encryption to prevent unauthorized access or breaches. Again, use AI to help you create these methods.
- **Compliance with Regulations:** Stay informed about data privacy regulations that apply to your business, such as GDPR (General Data Protection Regulation is a European Union regulation on information privacy in the European Union and the European Economic Area) or CCPA (California Consumer Privacy Act is a state statute intended to enhance privacy rights and consumer protection for residents of the state of California in the United States). Ensure that your AI practices comply with these regulations.

- **Data Minimization:** Collect and store only the data necessary for AI operations. Avoid collecting excessive or irrelevant data that could pose privacy risks.
- **Secure Data Handling:** Implement secure data handling practices, including access controls, data retention policies, and regular security assessments.

BUILDING TRUST WITH AI

- **Explainable AI:** Consider using concise AI models, allowing you to provide clear descriptions for AI-driven decisions when necessary.
- **Error Handling:** Be prepared to address AI errors and inaccuracies promptly. Transparently communicate when issues occur; outline and provide steps to rectify them.
- **User Education:** Explain how users will interact with AI systems effectively and what to expect. Providing guidelines and support can improve the user experience.
- **Accountability:** Assign accountability for AI systems within your organization and have supervising parties ensure that AI is used ethically and effectively.

Responsible AI usage is not only an ethical imperative but also a competitive advantage. It enhances your brand's reputation, fosters customer trust, and minimizes legal and regulatory risks. By following ethical guidelines, prioritizing data privacy and security, and building trust with AI, small businesses can fully harness the benefits of AI while maintaining their commitment to responsible usage.

CASE STUDY: PRIORITIZING ETHICAL AI PRACTICES

A small, environmentally-focused cleaning service in San Francisco prides itself on using sustainable practices and products. As they integrated AI into their operations for efficiency, they experienced a disconnect in using technology aligned with their ethical standards and commitment to customer privacy.

Challenge: The main goal for this small business was to employ AI tools to respect customer privacy, adhere to ethical standards, and maintain the trust they had built with their environmentally-conscious customer base.

Solution: They took several steps to ensure responsible and ethical AI usage and used the ChatGPT prompts below.

Prompts: Create clear and accessible information for customers explaining how AI is used in our operations and how we handle data. Also, include details in service agreements and create copy to post on our website about the use of AI, data collection, and privacy practices

Outcome: Listed information on their website and in service agreements detailing the use of AI and data handling practices. Customers appreciated the transparency, particularly regarding data collection, which increased trust and confidence in their services.

Prompt: Choose AI solutions with robust encryption and data protection policies to ensure customer information is secure.

Outcome: Selected AI tools that prioritized data security and privacy, ensuring customer information was protected. No reported instances of data breaches or privacy issues, preserving customer trust and safeguarding sensitive information.

Prompt: Conduct periodic audits of AI tools to check for biases in customer interaction and service delivery algorithms and provide recommendations to ensure fairness and lack of bias.

Outcome: Regularly reviewed their AI tools for biases, especially in customer interaction and service delivery algorithms, ensuring fair and unbiased AI interactions, contributing to equitable service delivery and customer satisfaction.

Prompt: Develop and conduct workshops for employees on ethical AI usage, focusing on privacy and responsible handling of AI-generated information. Also, create guidelines on how to handle AI-generated information and customer queries related to AI.

Outcome: Conducted employee workshops on responsible AI usage, emphasizing the importance of ethical practices and customer privacy. Employees became more confident and responsible in using AI, aligning their actions with the company's ethical standards.

Prompt: Set up an online feedback form and a dedicated contact point for AI-related inquiries and concerns. Create a system for customers and employees to report any concerns regarding AI usage, ensuring continuous improvement.

Outcome: Developed an online feedback form and a dedicated contact for AI-related inquiries and concerns. ChatGPT also provided a platform for continuous improvement based on customer and employee feedback, enhancing the ethical use of AI.

Prompt: Carefully select AI service providers that share our values of sustainability and ethical technology use [included values].

Outcome: Partnered with AI service providers that share their values of sustainability and ethical technology use. This strengthened their commitment to ethical practices, ensuring the alignment of AI.

Prompt: Monitor and adjust AI integration based on customer and employee feedback.

Outcome: Continuously monitored AI integration and made adjustments based on customer and employee feedback. AI systems remained aligned with user needs and ethical standards, facilitating ongoing improvement.

Results: After a year of implementation, this cleaning service experienced increased customer trust, operational efficiency, positive public perception, and employee engagement.

- **Maintained Customer Trust:** Customer feedback indicated a high level of trust in how they used AI, with no reports of privacy concerns.
- **Increased Operational Efficiency:** AI integration streamlined scheduling and customer communication, improving overall efficiency by 37.8%.
- **Positive Public Perception:** Their commitment to ethical AI usage enhanced their reputation, aligning with their brand as a sustainable, responsible business.
- **Employee Engagement:** Staff reported feeling more confident using AI tools, knowing that ethical practices were in place.

Conclusion: This case study demonstrates the importance of ethical considerations and responsible AI usage in small businesses. By prioritizing transparency, privacy, and fairness, they streamlined their operations with ChatGPT and strengthened their customers' and employees' trust and loyalty. This approach highlights how small businesses can responsibly integrate AI into their operations, ensuring that technological advancements align with their core values and ethical commitments.

ASK THE BOT: ADDITIONAL AI COMMANDS

To assist you in applying the knowledge you learned from this chapter, here are five prompts for small businesses on how to use ChatGPT to address ethical considerations and ensure responsible AI usage. This includes fair and honest practices, data privacy, security, and building trust. Open ChatGPT on your mobile device or computer and enter the following queries.

> What are the best practices for ensuring that [enter your business, website, or industry]'s use of ChatGPT and AI technologies aligns with ethical standards and promotes fairness in decision-making processes?

> How can [enter your business, website, or industry] implement robust data privacy and security measures when using ChatGPT to handle customer data and sensitive information, ensuring compliance with regulations and safeguarding user privacy?

> Building trust with our customers is extremely important to us. What strategies can we employ to develop and maintain trust when using ChatGPT and AI, especially when interacting with sensitive or confidential matters for [enter your business, website, or industry]?

> We're concerned about mitigating bias and discrimination in AI-driven interactions. What steps should we take at [enter your business, website, or industry] to identify and rectify biases in our ChatGPT responses, ensuring fairness and inclusivity?

> How can we maintain transparency and accountability in our AI usage, including ChatGPT? Are there tools or practices that can help us provide explanations for AI-driven decisions and actions for [enter your business, website, or industry]?

These inquiries will help small businesses navigate the ethical considerations associated with AI usage, promote responsible AI practices, and ensure that their use of ChatGPT aligns with ethical and privacy standards.

ADDITIONAL TRAINING

Since ChatGPT is continually evolving and each industry is unique, we have created extensive online training programs on our website, www.DynamicChatGPT.com, to further assist you in harnessing the power of AI for your small business.

CHAPTER 13
AI RESOURCES AND TOOLS FOR BUSINESSES

WE HAVE several recommendations for those who wish to dive deeper into AI and ChatGPT, including our online training courses at www.DynamicChatGPT.com. In this chapter, we'll explore the wealth of resources and tools available to help businesses harness the power of AI. Whether you're just starting your AI journey or looking to expand your AI capabilities, these resources can provide valuable guidance and support.

RECOMMENDED READING AND LEARNING MATERIALS

1. **Books:** Consider reading publications like *Artificial Intelligence: A Guide to Intelligent Systems* by Michael Negnevitsky, *Python Machine Learning* by Sebastian Raschka and Vahid Mirjalili, or *AI: A Very Short Introduction* by Margaret A. Boden. These books comprehensively introduce AI concepts, practical applications, and best practices. Since AI is such a quickly evolving technology, more books, including ours, are dropping on the market daily.
2. **Online Courses and MOOCs:** Platforms like Coursera, edX, and Udemy offer a variety of AI-related courses, including "Machine Learning" by Stanford University on Coursera

and "Deep Learning Specialization" on Coursera by Andrew Ng. Also, visit our website at www.DynamicChatGPT.com for in-depth online training on utilizing AI for small businesses. All these courses provide in-depth knowledge and hands-on experience.

3. **Blogs and Newsletters:** Stay updated with AI trends and insights by following AI-focused blogs like OpenAI's blog, *Towards Data Science* on Medium, *Dynamic ChatGPT*, and *AI News*. Subscribing to newsletters like the *MIT Technology Review*'s AI newsletter can also keep you informed.

4. **AI Podcasts:** Podcasts like "Artificial Intelligence with Lex Fridman," "Dynamic ChatGPT," and "TWiML & AI (This Week in Machine Learning & AI)" offer in-depth discussions with AI experts, providing valuable insights and real-world applications.

5. **AI Communities:** Engage with AI communities like Reddit's r/MachineLearning and AI Stack Exchange. These platforms provide opportunities to ask questions, share knowledge, and collaborate with AI enthusiasts and professionals.

AI DEVELOPMENT PLATFORMS AND TOOLS

- **OpenAI GPT Free and Paid Versions:** Currently, OpenAI's GPT-3.5 is free, while their GPT-4.0 and BETA versions have a minimal cost. These platforms offer powerful language models that can be fine-tuned for various applications, including chatbots, content generation, and more.

- **Grok, Bard, and Claude:** These are all specialized generative AI models. Grok excels in data interpretation and analysis, Bard focuses on creative content generation and storytelling, and Claude specializes in conversational AI and contextual understanding.

- **TensorFlow:** TensorFlow is an open-source machine learning framework developed by Google. It's widely used for building and training machine learning models, making it suitable for various AI applications.

- **PyTorch:** PyTorch is another popular open-source deep-learning framework known for its flexibility and ease of use. Researchers and developers for AI development particularly favor it.
- **Scikit-learn:** Scikit-learn is a Python library that provides simple and efficient data analysis and modeling tools. It's an excellent choice for small businesses looking to implement machine learning solutions.
- **Microsoft Azure AI:** Microsoft Azure offers a suite of AI services, including Azure Machine Learning, which provides tools for building, training, and deploying machine learning models.

SUPPORT AND COMMUNITY RESOURCES

- **AI Forums and Communities:** Participate in AI forums like Stack Overflow's AI section and join communities on platforms like GitHub to collaborate, seek help, and share knowledge with other AI practitioners.
- **Online Courses and Tutorials:** Explore AI-related courses on platforms like Coursera, Udacity, and edX, which often include forums and community support for learners.
- **AI Meetups and Conferences:** Attend local AI meetups or virtual conferences to network with other professionals and stay updated on the latest AI developments.
- **AI Consulting Services:** Consider consulting firms and AI experts who can provide personalized guidance and support tailored to your specific business needs.
- **Vendor Support:** If you use AI tools or platforms from vendors like Microsoft, Google, or AWS, take advantage of their support services and documentation.

These resources and tools can empower small businesses to navigate the AI landscape effectively. Whether you want to learn more about AI, develop AI solutions, or seek support and guidance from the AI community, these resources offer a valuable foundation for your AI journey.

ASK THE BOT: ADDITIONAL AI COMMANDS

To assist you in applying the knowledge you learned from this chapter, here are five inquiries for small businesses on how they can use ChatGPT resources and tools, including AI development platforms and tools, AI communities, and recommended reading and learning materials. Open up and log into ChatGPT on your mobile device or computer and enter the following queries.

What AI development platforms and tools are recommended for [enter your business, website, or industry] to integrate ChatGPT into their operations? What are the user-friendly options for companies with limited technical expertise?

How can [enter your business, website, or industry] tap into AI communities and forums to connect with experts and enthusiasts who can provide guidance and share tools for maximizing the potential of ChatGPT within our small business?

Suggest recommended reading materials, online courses, or tutorials to help us deepen our understanding of AI and ChatGPT and acquire the knowledge and skills necessary to successfully implement AI into our business [enter your business website, or industry].

Where can we find resources and guides tailored to training ChatGPT effectively, especially when customizing its responses to [enter your business, website]'s needs and [your industry]?

How can [your business, website] in [your industry] access reliable sources of information and stay informed about new features, updates, and evolving use cases for ChatGPT?

These inquiries will assist you and your small business access valuable resources, tools, and communities to facilitate their journey in using ChatGPT effectively and staying informed about AI advancements.

ADDITIONAL TRAINING

Since ChatGPT is continually evolving and each industry is unique, we have created extensive online training programs on our website, www.DynamicChatGPT.com, to further assist you in harnessing the power of AI for your small business.

CHAPTER 14
CONCLUSION

IMAGINE a world where the small business owner wakes up not to an overwhelming list of administrative tasks but to a summary of completed activities handled overnight by an AI assistant. Envision a scenario where customer inquiries from different time zones are not stuck in a bottleneck, but are now opportunities to showcase exceptional AI-powered 24/7 customer service. Picture a marketing strategy that evolves not through trial and error but through data-driven insights with campaigns crafted by AI to resonate with the target audience, delivering results previously only achievable by enterprises with deep pockets.

In this world, the small business owner wears fewer hats, each more splendid than the last, tailored by AI to fit perfectly. The hats of strategy, growth, and innovation are the new adornments of the entrepreneur who embraces AI. This is the future made possible by the democratization of technology and tools like ChatGPT that have been handed to us to forge a new path for small business prosperity.

The journey into AI is fraught with challenges but also ripe with opportunity. We have the capacity and technology to redefine what it means to be a small business in the digital age, to level the playing field, and to chart a course through the waters of technological change with confidence and clarity.

DISCOVERIES FROM USING CHATGPT IMPLEMENTATION

As we have seen from the valuable case studies at the end of each chapter, the uses and capabilities of ChatGPT implementation are endless, especially with small businesses. We have just scratched the surface. Below is a summary of the lessons learned.

1. **Tailored Solutions:** Small businesses should customize ChatGPT to address their unique needs. One size doesn't fit all, and tailoring AI solutions to specific challenges can lead to better outcomes.
2. **Efficient Use of Resources:** ChatGPT can help small businesses optimize resource allocation. Whether automating customer support, streamlining content creation, or enhancing e-commerce experiences, AI can efficiently use time and workforce to your advantage.
3. **User-Centric Approach:** Prioritize user experiences when integrating AI. Both customers and employees should find the AI interface intuitive and helpful.
4. **Continuous Improvement:** AI implementations require ongoing monitoring and fine-tuning. Regularly evaluate AI performance and gather user feedback to make necessary adjustments.
5. **Ethical Considerations:** Businesses should consider ethical concerns when using AI, such as data privacy, bias, and transparency. Addressing these issues proactively is essential for long-term success.
6. **Human-AI Collaboration:** Combining human expertise and AI capabilities often leads to the best results. In cases where AI handles routine tasks, human oversight and planned intervention are paramount for quality control.
7. **Return on Investment (ROI) and Scalability:** Small businesses should assess the ROI of their AI implementations. Additionally, consider scalability and ensure that AI solutions grow with your business.

The case studies and lessons learned highlight the versatility and potential of AI, particularly ChatGPT, in various business contexts. As AI technology advances, small businesses can leverage these tools to drive growth, improve efficiency, and remain competitive in today's rapidly evolving business landscape.

THE PATH AHEAD

As this book draws to a close, we look to the future with an amalgam of hope and anticipation and a blueprint for action. This is not the end but a commencement—a journey into the practicalities of AI, which, when wielded with finesse and ethical consideration, can elevate the small business owner from the mires of operational obscurity to the heights of competitive prowess and market relevance.

We have provided a manual of essentials, a companion for the modern entrepreneur who understands that the only constant in business is change, and the best way to predict the future is to create it. Throughout this primer, we've unraveled the intricacies of AI, provided a tactical approach for its adoption, and showcased how ChatGPT, a creation of OpenAI, stands as a beacon for small businesses navigating the digital transformation.

This journey began with an understanding that small business owners are the lifeblood of innovation but often find themselves shackled by the constraints of limited resources and time. The narrative has since shifted, with AI emerging as the great equalizer. ChatGPT and its contemporaries offer an arsenal of tools for customer service enhancement, data-driven decision-making, marketing and sales optimization, and automation of repetitive tasks.

We've learned that implementing AI must be strategic, aligned with the company's core values, and approached with a pragmatic mindset. We hope this resource has equipped you, the entrepreneur, with the knowledge to employ AI effectively, recognizing that with great power comes great responsibility—the responsibility to use AI ethically and in a manner that advances the business and the well-being of its customers and society. You have an incredible tool at your disposal and discretion.

Armed with this knowledge, the path ahead for small business owners is illuminated with the potential of AI. The landscape is one of dynamic evolution, where staying ahead of the curve is not merely an advantage but a necessity. Embracing AI technologies like ChatGPT can unlock untapped opportunities, foster innovation, and create a strategic advantage that allows small businesses to punch above their weight in the competitive arena.

The journey forward is not without its challenges. The rapid pace of technological change demands vigilance, continuous learning, and adaptability. Small businesses must stay informed of emerging AI trends, potential regulatory shifts, and the ever-evolving landscape of customer expectations. This book has laid the foundation, but the onus remains on each business leader to build upon and adapt to it.

PREPARING FOR THE HORIZON

Preparation is key as we stand on the brink of this new era. Entrepreneurs must invest in their knowledge and skills, embracing lifelong learning as a core tenet of their business strategy. Our vision for continuing education on AI extends beyond the pages of this book through the comprehensive training programs available on our website, www.DynamicChatGPT.com, and into the heart of each business venture willing to take the leap.

The future beckons with promise and potential pitfalls alike. It is a future where ChatGPT can be the silent partner in every small business —the digital ally in every entrepreneurial quest. This is a call to action, or rather a summons to engage with AI not as a distant concept but as a tangible, invaluable ally in the quest for growth, innovation, and sustainable success.

ASK THE BOT: ADDITIONAL AI COMMANDS

Here are your final five prompts for small businesses looking to learn from the lessons of integrating ChatGPT, focusing on tailoring solutions, efficient resource use, a user-centric approach, continuous improvement, and ethical considerations. Log into ChatGPT on your mobile device or computer and enter the following queries.

> What lessons can [enter your business, website, or industry] learn from companies that successfully integrated ChatGPT into their operations, particularly in tailoring AI solutions to meet their business needs and goals?
>
> What are the key takeaways from past integrations that can help us make the most efficient use of resources when implementing ChatGPT, ensuring a cost-effective approach to AI adoption for [enter your business, website, or industry]?
>
> How have businesses in [your industry] prioritized a user-centric approach when integrating ChatGPT, and what benefits have they experienced from focusing on delivering an exceptional user experience?
>
> What strategies and insights can we gain from others on the journey of integrating ChatGPT regarding the importance of continuous improvement, adapting to changing needs, and enhancing AI capabilities for [enter your business, website, or industry]?
>
> As we integrate ChatGPT into [enter your business, website, or industry]'s processes, what ethical considerations should we be aware of, and what safeguards can we put in place to ensure responsible AI usage and data privacy?

These inquiries will help small businesses learn from the experiences of others, adopt best practices, and make informed decisions when integrating ChatGPT into their operations.

ADDITIONAL TRAINING

Since ChatGPT is continually evolving and each industry is unique, we have created extensive online training programs on our website, www.DynamicChatGPT.com, to further assist you in harnessing the power of AI for your small business.

FINAL THOUGHTS

Enjoy the journey to the next phase of the digital revolution! In anticipation of great things, we are excited for you to implement AI into your business and industry. Email us your success stories at info@ DynamicChatGPT.com. Keep staying ahead of the curve!

EPILOGUE
[AT LEAST UNTIL THE NEXT UPDATE]

THE NEXT ERA: AI-DRIVEN SMALL BUSINESS OF THE FUTURE

DYNAMIC CHATGPT: *AI Strategies for Small Business* concludes not with a period but with an ellipse, an open invitation to continue the dialogue and to explore, question, and innovate. As this next era of technology is employed in the small business arena, know that we are committed to helping you, the small business owner, succeed. Remember to visit our website, www.DynamicChatGPT.com, for additional insights, recommendations, and training.

We have provided the map, but the terrain must be navigated by you, the astute business owners of today and tomorrow. It is your job to embrace AI to move your company ahead of the curve, utilizing ChatGPT, into the next era—the Artificial Intelligence revolution.

Here's to the human spirit that accesses the pulse of ChatGPT.

Stay inspired,

Ron and Kim

BONUS: USEFUL EVERYDAY COMMANDS FOR AI

IN THIS FUN BONUS SECTION, we'll delve into the world of fun, everyday user non-business commands for AI that have nothing to do with business because even entrepreneurs need a break from work!

ChatGPT is designed to assist users by providing information, answering questions, and engaging in natural language conversations. By understanding and utilizing common user commands, you can make the most of ChatGPT's capabilities for everyday life.

1. ASKING QUESTIONS

This is one of the primary ways to interact with ChatGPT. You can pose questions on various topics, from simple queries to complex inquiries. The AI will do its best to provide accurate and informative answers. Here are some examples.

> Applying General Knowledge: Query "How do I incorporate what Marie Curie did into our medical equipment business to increase sales?"

Concept Clarification: Seek explanations for concepts, such as "Explain how the greenhouse effect influences plastic products in my business line" or "What is blockchain technology?"

Definitions: Request definitions for words or phrases by asking, "Define biodiversity in my [insert my industry]'" or "What does 'onomatopoeia' mean and how is it used in ad creation?"

2. INFORMATION RETRIEVAL

Users can utilize ChatGPT as a resource to retrieve information on various subjects. Whether you need facts, statistics, or historical data, ChatGPT can help.

Historical Events: "Provide three examples of heroism on the battlefield during D-Day?" or "Who influenced the Renaissance artists the most and what was the result?"

Statistical Data: "Display a growth curve over the past fifty years for the Chinese economy." Ask for data and statistics, such as "What is the current population of China?" or "What's the GDP of Japan?"

3. MATHEMATICS AND CALCULATIONS

ChatGPT can perform mathematical calculations and provide mathematical information. Here are some examples of mathematical commands you can use.

Mathematical Theory and Calculations: "What were Euclid's influences in his contributions to geometry?" or "What is the square root of 12 multiplied by 7?" or "What is the difference between symbolic integrals and derivatives?"

> Complex Math: Get answers to more complex math questions, such as "What is the integral of x^2?" or "Solve the quadratic equation: 2x^2 - 5x + 3 = 0."

4. UNIT CONVERSIONS

ChatGPT can assist with unit conversions, making it a handy tool for tasks involving different units of measurement.

> Length and Distance: "How many 22-foot cables would be needed to cover a 5 km distance." Convert Units: "Convert 5 miles to kilometers" or "How many meters are in a mile?"

> Volume Conversion: "What is the impact per square foot of concrete if I dilute the recommended mixture by 1.2%?"

5. LANGUAGE TRANSLATION

Language translation is another powerful feature of ChatGPT. You can translate words, phrases, or sentences between different languages.

> Single Words or Phrases: "What is the formal Korean business greeting?" Translate words or short phrases, such as "Translate 'I need a taxi' into Spanish" or "What's the French slang phrase for 'How are you'?"

> Entire Sentences: Translate complete sentences like "Translate 'How are you?' into German" or "What does 'Je t'aime' mean in English?" Regional translation differences, such as "What is the difference between the Columbian and Mexican translation for 'I have to take my dog to the vet?'"

6. DATE AND TIME

ChatGPT can assist with questions related to date and time, including time zone conversions and historical events.

> Event and Time Inquiries: "What is the date and time of the next solar eclipse in the southern U.S. hemisphere?" Ask for the current date and time by saying, "What's the date today?" or "What time is it in Ndola, Zambia, right now?"
>
> Historical Dates: Inquire about historical events or dates, such as "What happened on July 20, 1969?" or "When was the Declaration of Independence signed?"

7. RECOMMENDATIONS

ChatGPT can recommend various topics, including books, movies, restaurants, and travel destinations.

> Book Recommendations: Seek reading suggestions with questions like "Can you recommend a good historical mystery novel?" or "What's a classic book everyone should read?"
>
> Movie and TV Show Recommendations: Ask for movie or TV show recommendations based on your preferences, such as "Recommend a comedy movie" or "What's a popular sci-fi series?"

8. CODE AND PROGRAMMING HELP

If you're a developer or have coding-related questions, ChatGPT can assist with programming tasks and explanations.

> Code Help: Request help with coding tasks by asking questions like "How do I write a Python program to calculate prime numbers?" or "Explain the 'if' statement in programming."

9. GEOGRAPHY AND TRAVEL

ChatGPT can provide information about geographical locations, travel tips, and directions.

> Geography Facts: Ask about geographical facts, such as "Tell me about the Great Wall of China" or "What are the major rivers in Africa?"

> Travel Directions: Get directions and advice, e.g., "Show me the most scenic route from Los Angeles to Denver." or "What are the must-visit places in Rome?"

10. HEALTH INFORMATION (DISCLAIMER REQUIRED)

While ChatGPT can provide general health-related information, it's vital to remember that it's *never* a substitute for professional medical advice. Always consult a healthcare professional for medical concerns.

11. ENTERTAINMENT AND POP CULTURE

For entertainment enthusiasts, ChatGPT can answer questions related to movies, music, celebrities, and more.

> Movie Plots: Ask about movie plots or details, such as "What is the plot of Inception?" or "Tell me what the latest Marvel movies' Easter eggs are."

> Actor Information: Inquire about actors and actresses, like "What other actors were considered for the lead role in 'The Shawshank Redemption?'" or "Tell me about Meryl Streep's method of acting."

12. EVERYDAY ADVICE

ChatGPT can provide advice and information on everyday topics, making life more fun and exciting.

> Cooking: Ask for recipes, cooking tips, or ingredient substitutions, such as "What's a good recipe for homemade pizza dough?" or "Can I substitute buttermilk for regular milk in this recipe?"

> DIY and Home Improvement: Seek advice on home improvement projects, like "How do I install a ceiling fan?" or "What tools do I need to paint a room?"

These common user commands for ChatGPT cover various topics and functionalities, allowing users to access information, perform calculations, and engage in natural conversations with this powerful AI language model. While ChatGPT is a valuable tool for information retrieval and assistance, it's essential to use critical judgment and verify information from reliable sources, especially for urgent or specialized matters. With these commands in your arsenal, you can maximize ChatGPT's capabilities and enhance your interactions with this AI assistant.

ADDITIONAL REFERENCES

For further reading and research on AI for small businesses, consider exploring the following references and citations:

1. Negnevitsky, M. (2019). *Artificial Intelligence: A Guide to Intelligent Systems*. Pearson.
2. Raschka, S., & Mirjalili, V. (2019). *Python Machine Learning*. Packt Publishing.
3. OpenAI Blog: [https://www.openai.com/blog/] (https://www.openai.com/blog/).
4. *Towards Data Science* (Medium publication): [https://towardsdatascience.com/] (https://towardsdatascience.com/).
5. *MIT Technology Review* AI newsletter: https://www.technologyreview.com/.
6. "AI with Lex Fridman" (Podcast): https://lexfridman.com/ai/.
7. "TWiML & AI (This Week in Machine Learning & AI)" (Podcast): https://twimlai.com/.
8. TensorFlow: [https://www.tensorflow.org/] (https://www.tensorflow.org/).
9. PyTorch: https://pytorch.org/.

10. Scikit-learn: https://scikit-learn.org/.
11. Microsoft Azure AI: https://azure.microsoft.com/en-us/services/ai/.
12. Stack Overflow (AI section): https://stackoverflow.com/questions/tagged/artificial-intelligence.
13. Artificial Intelligence through ChatGPT created the outline and provided the backbone of information and the documentation for this book: [OpenAI (2023). *ChatGPT* (September 25 Version) [Large language model](https://chat.openai.com).

These resources provide a wealth of information and guidance for small business owners looking to explore the world of AI and leverage its potential for growth and innovation.

———

ABOUT THE AUTHORS

RON CLARK

Ron Clark is a problem solver, solution creator, and visionary entrepreneur.

As the CEO of Clark Intl Group, he brings a wealth of experience in management and entrepreneurship to the table. Ron's entrepreneurial journey began over thirty-five years ago in the emerging deregulated field of telecommunications. Since then, he has created successful businesses in the IT, financial services, and energy deregulation fields.

With a Bachelor of Science in Finance from Florida Southern University, Ron has a solid educational foundation that complements his extensive practical knowledge. He has voraciously extended his knowledge base to include an Artificial Intelligence Certification from the University of Pennsylvania–Wharton School and an Artificial Intelligence and Machine Learning Certification from Google Labs. Currently, Ron is enrolled in the Mathematics and Statistical Modeling Program at the London School of Economics.

Throughout his career, he has excelled at recruiting and training exceptional talent while developing start-up companies. He has built sales and marketing teams of over 1,500 consultants internationally. His strategic acumen, organizational prowess, and business development expertise have been instrumental in navigating the complexities of the sales cycle and fostering business growth.

Ron's innovative spirit extends to his public speaking engagements, where he shares his insights on sales, marketing, and technical know-how. His outstanding communication skills and interpersonal aptitude have earned him numerous accolades and awards. He's been featured in and on the covers of multiple business magazines while leading substantial organizations. He has spoken to over 200,000 people around the country with audiences as large as 10,000.

Ron Clark's journey is a testament to his unwavering commitment to excellence and innovation in the world of business. His dedication to personal and professional growth has made him a true leader in the field. He thrives on coffee, giving belly-rubs to their 90-pound lab, and cutting-edge technology.

KIM M. CLARK, M.S.B.

Kim M. Clark is a seasoned marketing virtuoso, prolific author, and captivating communicator with over thirty-four years of corporate and entrepreneurial experience.

Kim is the dynamic Chief Marketing Officer (CMO) of Clark Intl Group. As an Amazon Bestselling author of six books (with the seventh en route) and a multiple award-winning speaker, she has proven her prowess in both the written and spoken word. With her visionary approach, a Master of Science in Marketing from Johns Hopkins University, and a Bachelor of Science in

Mass Communications from Towson State University, Kim's educational foundation has been instrumental in shaping her expertise.

Kim excels in creating and executing effective brand awareness and marketing strategies, setting the bar high for creative excellence. Her ability to create compelling narratives has earned her the prestigious AWSA Book of the Year Award and two Florida Tapestry Awards for Non-Fiction Books. She has also been a finalist for the Selah Literary Award, a testament to her literary talents.

In the realm of energy, Kim has achieved significant milestones with Ambit Energy, including the Pure Energy Award, membership in the Comma Club, and recognition as a 6-Figure Earner and Top Recruiter. She has been a recipient of two exclusive Five Star Trip awards, a testament to her outstanding achievements.

Kim's excellence extends to public speaking with her involvement with Toastmasters, where she has earned accolades such as the Best Speaker Award, Best Evaluator Award, and Best Table Topics Speaker Award. Her ability to captivate and inspire through her words is truly remarkable.

Beyond her written work and speaking engagements, Kim's influence is felt through her appearances in magazines like *South Orlando City Lifestyle* and *Victorious Living*. She has even graced the covers of two issues of *Success from Home Magazine*. Kim's voice has also resonated across the airwaves as she has been a frequent guest on radio, podcasts, and blogs, where she continues to inspire and inform audiences.

Kim M. Clark's career is a testament to her creative brilliance and unwavering dedication to marketing and communication. Her ability to inspire, educate, and innovate makes her a true powerhouse in her field, setting her apart as an industry leader. Kim is a genius in the kitchen, and her family goes into withdrawal when she's traveling for business.

―――――

AVAILABLE TRAININGS

For additional information, training, and consulting on leveraging ChatGPT for your business, scan the QR code below or visit us at www.DynamicChatGPT.com.